OSMOTIC DEHYDRATION & VACUUM IMPREGNATION

FOOD PRESERVATION TECHNOLOGY SERIES

Series Editor: Gustavo V. Barbosa-Cánovas

Innovations in Food Processing
    Editors: Gustavo V. Barbosa-Cánovas
             Grahame W. Gould

Trends in Food Engineering
    Editors: Jorge E. Lozano
             Cristina Añón
             Efrén Parada-Arias
             Gustavo V. Barbosa-Cánovas

Pulsed Electric Fields in Food Processing: Fundamental Aspects and Applications
    Editors: Gustavo V. Barbosa-Cánovas
             Q. Howard Zhang

Osmotic Dehydration and Vacuum Impregnation: Applications in Food Industries
    Editors: Pedro Fito
             Amparo Chiralt
             Jose M. Barat
             Walter E. L. Spiess
             Diana Behsnilian

FOOD PRESERVATION TECHNOLOGY SERIES

Osmotic Dehydration & Vacuum Impregnation

Applications in Food Industries

EDITED BY

Pedro Fito
Universidad Politécnica de Valencia

Amparo Chiralt
Universidad Politécnica de Valencia

Jose M. Barat
Universidad Politécnica de Valencia

Walter E. L. Spiess
Federal Research Centre for Nutrition, Karlsruhe

Diana Behsnilian
Federal Research Centre for Nutrition, Karlsruhe

Routledge
Taylor & Francis Group

LONDON AND NEW YORK

Osmotic Dehydration and Vacuum Impregnation

First published 2001 by Technomic Publishing Company

Published 2019 by Routledge
2 Park Square, Milton Park, Abingdon, Oxon OX14 4RN
52 Vanderbilt Avenue, New York, NY 10017

Routledge is an imprint of the Taylor & Francis Group, an informa business

First issued in paperback 2019

Main entry under title:
   Food Preservation Technology Series: Osmotic Dehydration and Vacuum Impregnation: Applications in Food Industries

Bibliography: p.
Includes index p. 243

Library of Congress Catalog Card No. 2001089548

ISBN 13: 978-0-367-45524-8 (pbk)
ISBN 13: 978-1-58716-043-1 (hbk)

Table of Contents

Part II: Vacuum Impregnation and Osmotic Processes in Fruit and Vegetables

Series Preface

W<small>E</small> welcome *Osmotic Dehydration and Vacuum Impregnation: Applications in Food Industries,* edited by Pedro Fito, Amparo Chiralt, Jose M. Barat, Walter E.L. Speiss and Diana Behsnilian to our fast growing Food Preservation Technology Series. This valuable addition to the Series covers an important topic never before treated in such a well-organized and comprehensive manner. This carefully edited reference addresses the fundamental and applied aspects of osmotic dehydration and vacuum impregnation and includes the work of highly renowned research centers from Europe, Israel, and Canada. I am particularly impressed with the depth of each chapter and the immense contribution of new knowledge this work brings to the scientific community and practitioners in food technology.

Those of us who have done work in this field are aware of the limitations to fully implementing this technology. There is no question that this book answers many of the previously unresolved issues and greatly facilitates the understanding and the scale-up of osmotic dehydration and vacuum impregnation.

I congratulate the editors and authors for a job well done. This is in an area in which a good book is long overdue. I hope all interested readers will experience as much enthusiasm and fun as I did going through the pages of this excellent and well thought out book.

GUSTAVO V. BARBOSA-CÁNOVAS
Series Editor

Preface

T HE principle of osmosis has been known as a means of water removal for some time. However, controlled application of osmotic treatments (OT) to food can be considered among the newest of improved techniques. Food products obtained for final consumption through OT are intermediate moisture products of improved quality, compared to conventionally dried materials. The treatment involves immersing foods in aqueous solutions of sufficiently high concentration at moderate temperatures. Consequently, water drains from the tissue into the solution and the solute transfers from the solution into the tissue. However, a leaching process of the tissue's own solutes into the solution is also observed.

OT is applied with the aim of modifying the composition of food material through partial water removal and impregnation, without affecting the material's integrity. A wide range of applications is possible through the appropriate choice and control of operating conditions, such as processing temperature, pressure and time, composition of solution, geometry of the food pieces, weight ratio solution, and contact between the food pieces and solution. The extent of water removal and solute impregnation is dependent on the additional processing techniques applied, and on the desired nutritional and sensory characteristics of the products.

The recent increase in interest in OT arises primarily from the need for quality improved food products. Quality improvement is related not only to water removal without thermal stress, but also to impregnated solutes. With the correct choice of solutes and a controlled and equilibrated ratio of water removal and impregnation, it is possible to enhance natural flavour and colour retention in fruit products, so that the addition of food additives such as antioxidants can be avoided; softer textures in partially dehydrated products can be obtained; and each food ingredient can be targeted for a particular use.

Due to the relatively simple equipment needed for batch operations, applications of OT have frequently neglected process optimisation; however, the development of industrial applications on a large scale demands a controlled process. There is much practical experience gained from OT alone, but to fulfil consumer, industrial, and environmental expectations, some problems remain to be solved.

For successful process control and optimisation, efforts must be made in the following key areas: (a) improvement in the understanding of the mechanisms of mass transport responsible for water removal and solute uptake, and achievement of a better insight into structural changes, so that the relationship between osmotic process variables and modifications achieved in the material can be used to develop predictive models; (b) prediction of the behavior of modified materials during further processing and storage; (c) response to environmental and economical questions for the management of osmotic solutions.

Adequate predictive models are needed to implement necessary process control and to achieve progress in the design of industrial equipment working in a continuous fashion. Consumers are interested in a wide range of safe products with excellent sensory and nutritional characteristics. Application of OT improves the overall quality of existing products and makes the development of new ones possible. However, optimization of the combined processing of foods, where osmotic dehydration is a step is still necessary. At the same time, management of osmotic solutions remains one of the most critical points on an industrial scale to be resolved.

This book includes edited and expanded versions of the papers presented at the "3rd Industrial Seminar on Osmotic Dehydration and Vacuum Impregnation: Applications of New Technologies to Traditional Food Industry", which took place March 15th, 1999 in Valencia, Spain at the Universidad Politécnica de Valencia. This Seminar Series was part of a European Union Concerted Action funded by the Directorate General XII (research grant FAIR 96-1118). Prior Seminars, part of this Series, were held in Porto, Portugal (October 1997) and Bertinoro, Italy (April 1998).

This EU Concerted Action involved 13 Research Centers and Universities in Europe (9 countries), Israel, and Canada, including: Federal Research Centre for Nutrition and University of Karlsruhe, Germany; CIRAD-AMIS, France; The Robert Gordon University, Scotland; Aristotelean University of Thessaloniki, Greece; University of Udine and I.V.P.T.A., Italy; ATO-DLO Institute, The Netherlands, Israel Institute of Technology, Israel; and University of Guelph, Canada. Two other European Institutions were associated with the group: Warsaw Agricultural University, Poland and The National Food Centre, Ireland.

The main objectives of this action were to create and improve the links between the different groups working on applications of osmotic treatments to food material, improve scientific knowledge for the evaluation and control of modifications of food processed through osmotic treatments, and provide necessary scientific and technological tools for industrial application of OT on a larger scale.

We hope this book will fill a wide gap in the understanding and utilization of osmotic dehydration and vacuum impregnation in food processing. We thank all scientists, institutions, and sponsors for making this work possible.

THE EDITORS

Acknowledgement

THE Editors wish to thank the Comisión Interministerial de Ciencia y Tecnología (Spain), FAIR Program (E.U, DGXII) and CYTED program for their financial support.

List of Contributors

BYRNE, H.
NESVADBA, P.
HASTINGS, R.
Food Science and Technology
  Research Centre
The Robert Gordon University,
  School of Applied Sciences
St Andrew Street, Aberdeen
Scotland, UK, AB25 1HG

GARRIDO, A.
BRENES, M.
GARCÍA-GARCÍA, P.
ROMERO, C.
Departamento de Biotecnología de
  Alimentos
Instituto de la Grasa
41012, Sevilla, Spain

MAVROUDIS, N.E.
LEE, K.-M.
SJÖHOLM, I.
HALLSTRÖM, B.
Food Engineering, Centre for
  Chemistry and Chemical
  Engineering
Lund University, Box 124, 221 00
Lund, Sweden

SHI, J.X.
Food Research Center, Agriculture
  and Agri-Food Canada
Guelph, Ontario N1G 2C9
Canada

LE MAGUER, M.
Department of Food Science
University of Guelph
Guelph, Ontario N1G 2W1
Canada

LAZARIDES, H.
Department of Food Science &
  Technology
Aristotelian University of
  Thessaloniki
Thessaloniki, Greece

TORREGGIANI, D.
BERTOLO, G.
I.V.T.P.A., Via Venezian 26
20133 Milano, Italy

ERLE, U.
SCHUBERT, H.
Institute of Food Process
  Engineering
Technical University of Karlsruhe
Karlsruhe, Germany

TORRINGA, E.
LOURENCO, F.
SCHEEWE, I.
BARTELS, P.
ATO-DLO Agrotechnological
  Research Institute
P.O. Box 17, 6700 AA
  Wageningen
The Netherlands

BRENNAN, M.H.
GORMLEY, T.R.
Teagasc, The National Food Centre
Castleknock
Dublin 15, Ireland

DABROWSKA, R.
LENART, A.
Department of Food Engineering
  and Process Management
Faculty of Food Technology
Warsaw Agricultural University,
  SGGW
166 Nowoursynowska St
02-787 Warsaw, Poland

ANDRÉS, A.
ALBORS, A.
ARGÜELLES, A.
BARAT, J.M.
CAMACHO, M.M.
CHÁFER, M.
CHIRALT, A.
ESCRICHE, I.
FITO, P.
GONZÁLEZ-MARTÍNEZ, C.
GRAU, R.
MARTÍN, M.E.
MARTÍNEZ-MONZÓ, J.
MARTÍNEZ-NAVARRETE, N.
ORTOLÁ, M.D.
PÉREZ-JUAN, M.
RODRÍGUEZ-BARONA, S.
SERRA, J.A.
Departamento de Tecnología de
  Alimentos
Universidad Politécnica de
  Valencia
Camino de Vera s/n
Apdo Correos 22012
46022, Valencia, Spain

MONTERO, A.
Departamento de Ciencia Animal
Universidad Politécnica de
  Valencia
Camino de Vera s/n
Apdo Correos 22012
46022, Valencia, Spain

CATALÁ, J.M.
DE LOS REYES, E.
Departamento de Comunicaciones
Universidad Politécnica de
  Valencia
Camino de Vera s/n
Apdo Correos 22012
46022, Valencia, Spain

GONZÁLEZ-MARIÑO, G.
Facultad de Ingeniería
Universidad de la Sabana
Campus Puente del Común
Chía Cundinamarca
Colombia

PAVIA, M.
FERRAGUT, V.
GUAMIS, B.
Departament de Patología i
    Producció Animal
Facultat de Veterinària
Universitat Autónoma de
    Barcelona
Edifici V 08193
Bellaterra
Barcelona, Spain

SALVATORI, D.
Departamento de Industrias
Facultad de Ciencias Exactas y
    Naturales
Ciudad universitaria
1428 Buenos Aires, Argentina

ANDÚJAR, G.
Food Industry Research Institute
Ave. R. Boyeros
Havana 13400, Cuba

BUGUEÑO, G.
Departamento de Agroindustrias
Universidad del Bio-Bio
Avda Andrés Bello s/n
Casilla 447
Chillan, Chile

PART I

OSMOTIC PROCESSES IN FRUIT AND VEGETABLES AT ATMOSPHERIC PRESSURE

CHAPTER 1

High-Quality Fruit and Vegetable Products Using Combined Processes

D. TORREGGIANI
G. BERTOLO

INTRODUCTION

CONSUMER demand has increased for processed products that keep more of their original characteristics. In industrial terms, this requires the development of operations that minimize the adverse effects of processing. At the moment there is a high demand for high-quality fruit ingredients to be used in many food formulations such as pastry and confectionery products, ice cream, frozen desserts and sweets, fruit salads, cheese, and yogurt. For such uses, fruit pieces must preserve their natural flavor and color, they must preferably be free of preservatives, and their texture must be agreeable. Proper application of "combined processes" may fulfill these specific requirements. These processes use a sequence of technological steps to achieve controlled changes of the original properties of the raw material (Maltini et al., 1993). In most cases, the aim is to obtain ingredients suitable for a wide range of food formulations, although end products can also be prepared. Although some treatments, e.g., blanching, pasteurization, and freezing, have primarily a stabilizing effect, other steps, namely, partial dehydration and osmodehydration, allow the properties of the material to be modified. Modification may include physical properties such as water content, water activity, and consistency, and chemical and sensory properties as well; the latter two are also associated with a change in composition. A partial dehydration step is useful to set the ingredients in the required moisture range, whereas a finer adjustment of water activity, consistency, sensory properties, and other functional properties is better achieved by what is generally called an "osmotic step," i.e., a temporary dipping in a concentrated syrup. As is now well recognized, under this conventional term of osmosis, there is a more complex phenomenon that has been defined as "Dewatering-Impregnation-Soaking" in concentrated solutions. According to temperature, time, type of syrup, and surface and mass ratio of product to solution, the osmotic treatment can induce on the same raw material very different effects. Of primary importance are the ratio of water loss, mainly affecting consistency, to solute uptake from the syrup, affecting flavor and taste.

Despite the large number of theoretical and experimental studies, combined processes are still hard to find in the food industry, although there are some "confidential" applications. The general criteria, here outlined, could be a first help in the choice and management of the proper technology. As an example, results are summarized of researches conducted to evaluate the effect of both raw material characteristics and process parameters on texture, color, and vitamin C retention in frozen kiwifruit slices after processing and during frozen storage.

RESULTS AND DISCUSSION

The first goal of the research was to obtain high-quality, high-moisture frozen kiwifruit slices for the preparation of frozen fruit salads. As a first step, the influence was studied of the ripening stage, and thus of the texture level of the raw fruit on the texture characteristics of freeze-thawed kiwifruit slices. Because there is a strong correlation between texture characteristics and pectic composition of fruits (Souty and Jacquemin, 1976), the different forms of pectins were analyzed on the basis of their different solubility: water soluble, oxalate soluble, and residual protopectin.

Kiwifruits, cultivar Hayward, were stored differently to obtain the following three texture levels: 4–5 kg (firm), 1.8–2.5 kg (medium), and 0.8–1.5 kg (soft). Osmotic dehydration in a 70% sucrose solution for 4 h at room temperature was used as a pretreatment considering the high sensitivity of kiwifruit color to air drying (Forni et al., 1990). The osmodehydrated fruits were then frozen and stored at −20°C for 12 months. As for the osmotic step, the lower the texture level the lower the soluble solids gain, whereas the water loss value is higher in the firm fruits compared with that of medium and soft ones (Torreggiani et al., 1998a).

A slight increase of texture was observed in the osmodehydrated kiwifruits, which could be due to the simultaneous loss of water and gain of solids caused by the process (Figure 1.1). Firm, medium, and soft kiwifruit show percentage texture increase of 6%, 13%, and 17%, respectively.

As shown in Figure 1.2, a significative decrease of texture due to the freezing process was observed (Torreggiani et al., 1998b). Firm, medium, and soft kiwifruit showed percentage texture decrease of 42%, 44%, and 62%, respectively, indicating a very poor suitability of the soft fruits. During the first 4

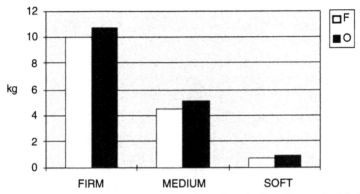

Figure 1.1 Texture values of kiwifruit slices before (F) and after (O) osmotic dehydration (Torreggiani et al., 1998a).

months of storage, texture of firm and medium fruits showed a further significant decrease. At the end of the 12 months of storage, the texture of the firm kiwifruits was still significantly higher than that of the medium and soft ones.

The analysis of the pectin composition confirmed the major role of the protopectin in tissue firmness (Shewfelt and Smit, 1972), as shown in Figure 1.3.

Furthermore, residual protopectin (R) degradation observed during freezing in firm and medium kiwifruit (Figure 1.3) could be regarded as one of the causes of the texture reduction observed in the freeze-thawed fruits. A direct correlation ($R^2 = 0.92$) was found between texture of the raw fruit and texture acceptance of the processed fruit all along the storage period (Figure 1.4), so indicating that the ripening stage of the raw fruit is a key point in the production of high-quality frozen kiwifruit slices.

The second step in the combined process optimization was to evaluate the influence of the osmotic solution on the stability of chlorophyll pigments and

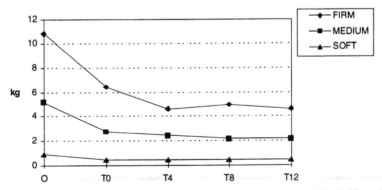

Figure 1.2 Texture values of kiwifruit slices after freezing (*T0*) and after 4 (*T4*), 8 (*T8*), and 12 (*T12*) months of storage at −20°C (Torreggiani et al., 1998b).

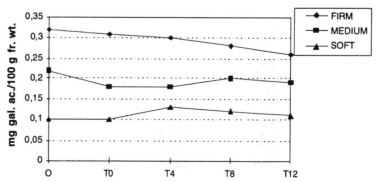

Figure 1.3 Pectin fraction R of kiwifruit slices after freezing (*T*0) and 4 (*T*4), 8 (*T*8), and 12 (*T*12) months of storage (Torreggiani et al., 1998b).

ascorbic acid content of osmodehydrofrozen kiwifruit slices during frozen storage. According to the kinetic interpretation based on the glass transition concept, chemical and physical stability is related to the viscosity and molecular mobility of the unfrozen phase, which, in turn, depends on the glass transition temperature (Levine and Slade, 1988; Roos, 1992). So it has been hypothesized that the cryostabilization of frozen foods depends on the ability to store the food at temperatures less than its glass transition (T_g) or the ability to modify the food formulation to increase glass transition temperatures above normal storage temperature (Slade and Levine, 1991). Kiwifruit slices 10 mm thick were osmodehydrated in a 60% solution of sorbitol, sucrose, and maltose for 4 h, and then were frozen and stored at $-10°C$, $-20°C$, and $-30°C$ for 9 months. The osmotic pretreatment modified the percent distribution of the sugars and thus the glass transition temperature of the fruits (Torreggiani et al., 1994). There was a correlation between the storage temperature and chlorophyll retention: the lower the temperature the higher the retention (Figure 1.5). At the same storage temperature, kiwifruit pretreated in maltose and

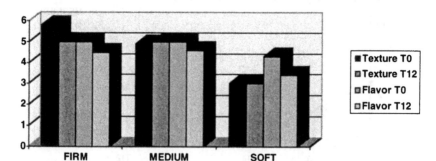

Figure 1.4 Texture and flavor acceptance of osmodehydrofrozen kiwifruit (Torreggiani et al., 1998b).

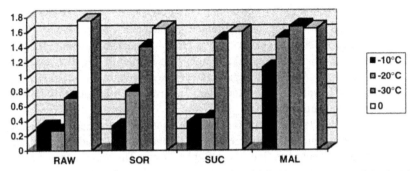

Figure 1.5 Chlorophyll content (mg/100 g fr. wt.) of kiwifruit slices, not pretreated (raw) and pretreated in sorbitol (SOR), sucrose (SUC), and maltose (MAL), after 9 months of frozen storage; 0 = content before freezing.

thus having the highest T_g values showed the highest chlorophyll retention. Ascorbic acid content was stable at $-20°C$ and $-30°C$, whereas there was a significative decrease at $-10°C$ (Figure 1.6). At $-10°C$ the ascorbic acid content showed the highest retention in the kiwifruit pretreated in maltose.

In the case of both chlorophyll and ascorbic acid content, the glass transition theory holds. Increasing the glass transition temperature through an osmotic step could increase the fruit stability during frozen storage. These data were confirmed by the results obtained by studying anthocyanin stability in osmodehydrofrozen strawberry halves (Forni et al., 1997a) and color and vitamin C retention in osmodehydrofrozen apricot cubes (Forni et al., 1997b). The protective effect of sorbitol observed both in osmodehydrated strawberry halves and apricot cubes could not be explained through the glass transition theory. Further research is needed to define all the different factors such as pH, viscosity, water content, and specific properties and characteristics of this sugar-alcohol, influencing pigment degradation.

The second goal of the research was to obtain high-quality, low-moisture kiwifruit products. Besides studying the dehydration methods (osmotic dehy-

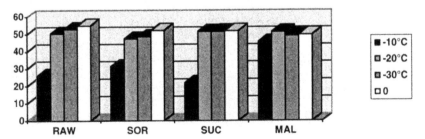

Figure 1.6 Ascorbic acid content (mg/100 g fr. wt.) of kiwifruit slices, not pretreated (raw) and pretreated in sorbitol (SOR), sucrose (SUC), and maltose (MAL), after 9 months of frozen storage; 0 = content before freezing.

dration, air dehydration, and their combination), the dehydration levels (20, 30, 40, 50, and 60% weight loss) and their influence on color of the end products were also studied. Up to 40% weight loss osmodehydration and its combination with air dehydration did not significantly modify the kiwifruit color, whereas air drying caused a significant yellowing (Figure 1.7). The results show that 40% weight loss is the limit for kiwifruit pretreatments if high-quality products are required.

By referring to this research on kiwifruit, it is easy to understand that despite the large number of theoretical and experimental studies, combined processes are still hard to find in the food industry. The wide range of possibilities regarding raw material characteristics, product characteristics, process parameters, and flow sheets make it very difficult for these processes to be applied. Because combined processes could be effective and useful tools for formulating new fruit ingredients with functional, sensory, and nutritional characteristics suitable for specific industrial uses, further research is needed to implement such interesting techniques.

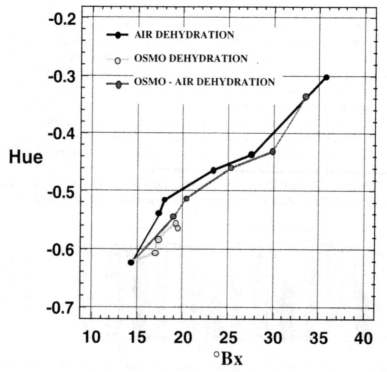

Figure 1.7 Hue values versus °Bx of differently prehydrated kiwifruit slices after thawing-rehydration to the soluble solid content of the initial fresh kiwifruit (~14°Bx).

REFERENCES

Forni, E., Genna, A., and Torreggiani, D. 1997a. Modificazione della temperatura di transizione vetrosa mediante disidratazione osmotica e stabilità al congelamento del colore delle fragole. In *Ricerche e innovazioni nell'industria alimentare, 3° Congresso Italiano di Scienza e Tecnologia degli Alimenti (CISETA 3)*, S. Porretta, Ed., Chiriotti Editori, Pinerolo (I), pp. 123–130.

Forni, E., Sormani, A., Scalise, S., and Torreggiani, D. 1997b. The influence of sugar composition on the colour stability of osmodehydrofrozen intermediate moisture apricots. *Food Res. Int.* 30(2):87–94.

Forni, E., Torreggiani, D., Crivelli, G., Maestrelli, A., Bertolo, G., and Santelli, F. 1990. Influence of osmosis time on the quality of dehydrofrozen kiwifruit. In *Acta Hort. 282*, A. R. Ferguson, Ed., ISHS, Wageningen, The Netherlands, pp. 425–434.

Levine, H. and Slade, L. 1988. Principles of 'Cryostabilization' technology from structure/property relationships of carbohydrate/water systems—a review. *Cryo Lett.* 9:21–63.

Maltini, E., Torreggiani, D., Rondo Brovetto, B., and Bertolo, G. 1993. Functional properties of reduced moisture fruits as ingredients in food systems. *Food Res. Int.* 26:413–419.

Roos, H. Y. 1992. Phase transitions and transformations in food systems. In *Handbook of Food Engineering*, D. R. Heldman and D. B. Lund, Eds., Marcel Dekker Inc., New York, pp. 145–197.

Shewfelt, A. L. and Smit, C. J. B. 1972. An estimate of the relationship between firmness and soluble pectin of individual peaches during ripening. *Lebensm. Wiss. Technol.* 5(5):175–177.

Slade, L. and Levine, H. 1991. Beyond water activity: Recent advances based on an alternative approach to the assessment of food quality and safety. *Crit. Rev. Food Sci. Nutr.* 30:115–360.

Souty, M. and Jacquemin, G. 1976. Dègradation de la texture des fruits appertisés au syrop. Etude sur l'hydrolyse de la protopectine des abricots. *Ind. Alim. Agric.* 2:39–15.

Torreggiani, D., Forni, E., and Pelliccioni, L. 1994. Modificazione della temperatura di transizione vetrosa mediante disidratazione osmotica e stabilità al congelamento del colore di kiwi. In *Ricerche e innovazioni nell'industria alimentare, 1° Congresso Italiano di Scienza e Tecnologia degli Alimenti (CISETA 1)*, S. Porretta, Ed., Chiriotti Editori, Pinerolo (I), pp. 621–630.

Torreggiani, D., Forni, E., Maestrelli, A., and Quadri, F. 1998a. Influence of osmotic dehydration on texture and pectic composition of kiwifruit slices. In *Proceedings 11th International Drying Symposium (IDS'98)*. Halkidiki, Greece, August 19–22, Vol. A, pp. 930–937.

Torreggiani, D., Forni, E., Mastrelli, A., and Bertolo, G., 1998b. Osmodehydrofreezing to improve frozen kiwifruit quality: The influence of raw fruit texture. In *Proceedings 3rd Karlsruhe Nutrition Symposium, European Research towards Safer and Better Foods*, V. Gaukel and W. L. E. Spiess, Eds., Part 2, pp. 353–362.

CHAPTER 2

Osmotic Treatment of Apples: Cell Death and Some Criteria for the Selection of Suitable Apple Varieties for Industrial Processing

N. E. MAVROUDIS
K.-M. LEE
I. SJÖHOLM
B. HALLSTRÖM

INTRODUCTION

OSMOTIC dehydration involves the partial dehydration of water-rich solid foodstuffs, through immersion in hypertonic aqueous solutions of various edible solutes. Research in the area of osmotic dehydration has been ongoing for more than 25 years, but the industrial application of the method still is limited for various reasons (Raoult-Wack, 1994). One of (if not) the most important reason remains that the structural complexity of the cellular tissue has been an impediment to getting a clear picture of the controlling mass transport mechanism(s) (Le Maguer, 1997). Extending the meaning of structure to include not only the skeleton of the foodstuff but the condition of the cells is of particular importance given that the cell "phase" constitutes by far the most voluminous "phase" in plant tissue. The availability of the cell "phase" for diffusion to a non-permeable solute such as sucrose is directly related to changes in membrane selectivity. Such changes can be a result of senescence and/or processing parameters, including temperature, pressure, osmotic medium concentration, etc. A way to evaluate the effect of aging on the membrane selectivity is by evaluating the ion leakage.

Marcotte and Le Maguer (1992) have claimed the possible existence of mechanisms other than diffusion for solute transport since early 1990s. Mavroudis et al. (1998a) reported the simultaneous existence of convection and diffusion phenomena during the osmotic treatment of apple fruit tissue at 20°C. As a result of these findings, two main questions were raised: (1) what is the significance of the convection phenomena on water loss and solid gain and (2) where do non-permeable solutes such as sucrose diffuse. Mavroudis

11

et al. (1998c) studied the cell viability when individual apple cells were immersed in 40% and 53% sucrose solutions at ambient temperature. By their experimental observations was assumed that at least the first few layers of cells die after immersion because of severe osmotic shock. However, the fact that only individual apple cells were used is a limiting factor because single cells could be of lower strength than cells in tissue pieces. In a subsequent research effort by the same group (Mavroudis et al., 1998b), consideration of the convection phenomena, together with cell death, seemed to explain the kinetics observed in the two apple varieties (Kim, Sweden, and Granny Smith, Argentina). In that study, the osmotic medium used was 50% sucrose solution, and experiments were performed at three different process temperatures, namely, 5, 20, and 40°C.

The main objectives of the present study were to:

(1) Evaluate the significance of the intercellular space accessibility in osmotic treatments by measuring the ability of apple fruit tissue to absorb apple juice.
(2) Investigate the possible influence of membrane selectivity on the kinetics of osmotic treatments, as is determined by electrical conductivity measurements.
(3) Evaluate the loss of cell vitality in actual experimental samples during immersion in 50% w/w sucrose solution at 20°C.

MATERIALS AND METHODS

SAMPLE PREPARATION—PROCESS CONDITIONS

Apple tissue (Kim and Mutsu variety, south Sweden) was used in cylindrical shape. The samples were taken by means of a cork borer and were cut equatorially from the surface into the core to sample the parenchymatic tissue between the seeds sacs. From each radial cylinder cutting at the point where the vascular tissue is at its densest, thus obtaining samples from the center and from the periphery of the parenchyma called inner and outer specimens, respectively, produced two apple samples. Those samples differ mainly in the tissue structure, i.e., cell and intercellular spaces arrangement and morphology (Mavroudis et al., 1998a). The osmotic dehydration experiments were performed in a jacketed, agitated vessel. The osmotic medium was 50% sucrose solution at temperature 5, 20, and 40°C, and the agitation was set to 4500 Re (refer to the sample).

For the evaluation of the significance of the convection phenomena, the osmotic medium was replaced by apple juice, and the experiments were performed either in the same agitated vessel at 4500 Re or using rotated tubes at 16000 Re (refer to the sample) held at ambient temperature.

The evaluation of the electrical conductivity (EC) was made by placing each apple specimen in a tube together with 30 ml of 3% sucrose solution. The specimen was left to rotate in 50 rpm for 60 min; then the surrounding liquid was collected and its electrical conductivity was measured.

The evaluation of the loss of cell vitality in actual experimental samples was made during immersion in 50% w/w sucrose solution at 20°C. The apples were Granny Smith variety. The samples were divided into control and treated groups. The control samples immediately after cutting were immersed in a triphenyltetrazoliumchloride (TTC) solution and placed under vacuum until the pores of the tissue were filled with the TTC solution and were left overnight in complete darkness. The same procedure was performed for the osmotically treated samples after the osmotic step. After the incubation period, the treated samples and controls were photographed under a stereomicroscope. The reduction of TTC in the mitochondria of the live cells produces a red water insoluble dye. Therefore, areas rich in alive cells in the tissue are stained dark red, whereas light colored areas are indicative of damaged and dead cells.

Further details of specimens' handling, formation, and the experimental procedure are given in Mavroudis et al. (1998a).

ANALYTICAL METHODS, CALCULATIONS AND STATISTICS

The EC was measured by using a CDM210 conductivity meter from Radiometer Copenhagen, France, and expressed at 25°C. The EC of the specimen was calculated by subtracting the initial EC from the final EC value of the 3% sucrose solution that was used as the soaking medium. The moisture content was measured before and after treatment by the freeze-drying method. The water loss and solid gain were expressed in g/g i.m. (initial matter) to account for initial weight differences between samples. Statistical analysis was performed by using STATGRAPHICS. The Multiple Regression Analysis module was used to elucidate the effect of EC on the process responses.

RESULTS AND DISCUSSION

INTERCELLULAR SPACE ACCESSIBILITY

The evolution of weight increase with time for both varieties, both kinds of cellular tissue, and for the two different experimental setups are shown in Figure 2.1. As it is apparent from this figure, in all cases the immersion of apple tissue in apple juice caused a weight increase in the tissue. A diffusive mechanism cannot account for such an increase, given that the concentration differences between the apple tissue and its juice could be considered negligi-

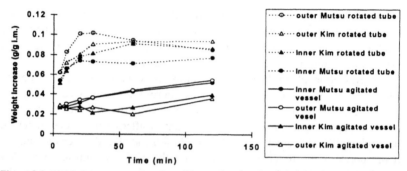

Figure 2.1 Weight increase as a function of immersion time in apple juice for two different experimental setups. Solid lines, agitated vessel; dashed lines, rotated tube. Open and filled symbols, outer and inner specimens, respectively.

ble. What most likely has happened is penetration of the apple juice into the intercellular spaces. This penetration could come as a result of convection phenomena, i.e., free and/or forced convection. Because in both of our experimental setups we have not had any of our phases (apple tissue and medium) stagnant it is likely for both mechanisms to appear. However, the forced convection is expected more pronounced in the rotated tube configuration (i.e., due to higher Re number) something that can explain the by far higher weight increase in that setup (Figure 2.1). In the case where the free convection is more pronounced (agitated vessel), we could see (Figure 2.1) a strong segregation between the two apple varieties, reflecting most probably their differences in the intercellular space accessibility. The latter can play a significant role not only between varieties but also between tissues of the same variety with different cell arrangement (Figure 2.1). It seems logical to assume that the replacement of apple juice by an osmotic medium will not prevent the uptake of the solution. Therefore, the convection phenomena could play a significant role in osmotic dehydration kinetics because they could influence both water loss and solid gain. Consequently, the evaluation of the intercellular space accessibility of an apple tissue seems to be a significant raw material characteristic for the prediction of the osmotic processing responses.

OSMOTIC TREATMENT KINETICS AND MEMBRANE SELECTIVITY

Figures 2.2(a), (b), and (c) show the evolution of solid gain and water loss as a function of time for both apple varieties. In both varieties the inner samples lost less water than the outer samples. Interestingly, after the first 30 min of processing, the samples from Mutsu variety (solid lines, Figure 2) were consistently of lower water loss than the samples from Kim variety. This coincides very well with the finding from the apple juice experiments (solid

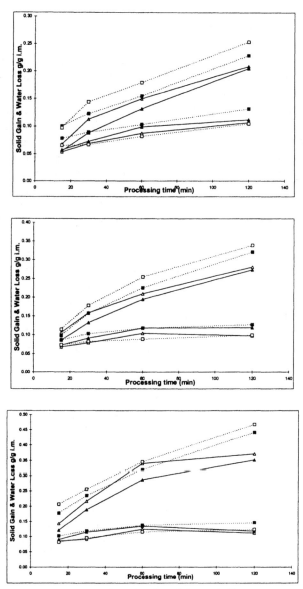

Figure 2.2 Solid gain and water loss as a function of time and all process temperatures. Solid lines Mutsu var., dashed lines Kim var. Open and filled symbols, outer and inner specimens respectively; (a) 5°C, (b) 20°C, (c) 40°C.

lines, Figure 2.1) where the absorption of apple juice between the two varieties is roughly similar in the beginning and increases thereafter. In the solid
gain, there is a strong separation between the inner and outer specimens rather
than between the two apple varieties. To explain such a finding, we need to
consider the simultaneous action of the convection phenomena and diffusion
as the two mechanisms that drive the solute uptake (Mavroudis et al., 1998a,
b, c). The diffusion of the solute (sucrose) is taking place (1) on the cell wall
area, which is hydrated, and (2) mainly in the first layer of cells under the
sample-medium interface only after the osmotically induced cell death of
those cells (Mavroudis et al., 1998a, b). Based on that, it seems possible the
cells from the inner specimens, which are elongated and double in volume
compared with that in the outer specimens (Mavroudis et al., 1998a), to
be more sensitive to the osmotic shock and thus resulting in higher solute
uptake.

In Table 2.1 the results are shown of a multiple linear regression on which
all data were subjected to reveal the influence of the membrane selectivity as
determined by the electrical conductivity measurement. As it is apparent (Table
1), the EC is a significant factor for the solid gain while not a proven significant factor for the water loss (analysis not shown). The higher the EC the
higher the solid gain is expected to be. An explanation could lie in the fact
that the increased EC reflects the low physical condition of the cells and the
latter means higher susceptibility to cell death.

TABLE 2.1. Multiple Regression Analysis Summary:
Dependent Variable, Solid Gain (g/g i.m.).

Estimated coefficients for the independent variables included in the model				
Parameter	Coefficient	Std Error	T Statistic	P Value
Variety	0.00534124	0.00132198	4.04033	>0.0001
Electrical conductivity	0.000382934	0.0000805704	4.75279	>0.0001
Initial water content	0.00642065	0.000698016	9.19843	>0.0001
Structure	−0.0166114	0.00155472	−10.6845	>0.0001
Temperature	0.000684535	0.000056892	12.0322	>0.0001
Square root of time	0.00575463	0.000288722	19.9314	>0.0001
Analysis of variance of the model				
Source	Sum Squares	of D.F.	Mean Square	F Ratio	P Value
Model	2.86006	6	0.476676	2728.48	>0.0001
Residual	0.0487424	279	0.0000175		
Total (Corr.)	2.9088	285			

R^2 = 0.983; R^2 (adjusted for DF) = 0.982; standard error of est. = 0.0132; mean absolute error = 0.0092.

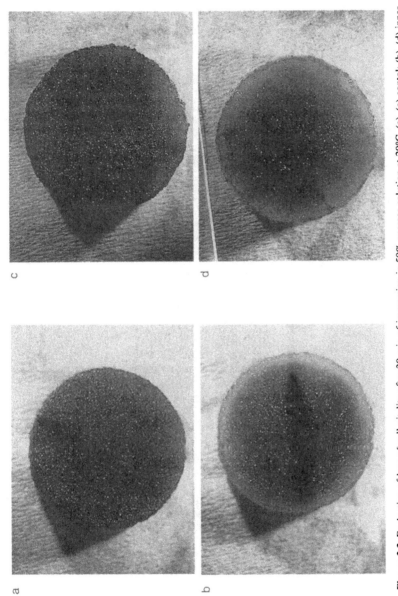

Figure 2.3 Evaluation of loss of cell vitality after 30 min of immersion in 50% sucrose solution at 20°C. (a), (c) contol; (b), (d) inner and outer treated samples, respectively.

17

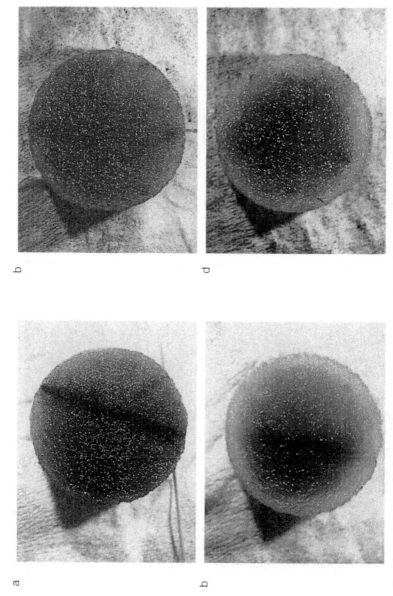

Figure 2.4 Evaluation of loss of cell vitality after 90 min of immersion in 50% sucrose solution at 20°C. (a), (c) contol; (b), (d) inner and outer treated samples, respectively.

LOSS OF CELL VITALITY IN THE ACTUAL SAMPLES

The osmotically treated samples, inners and outers, gave a light-colored layer of cells, starting on the surface of the samples and continuing in a thickness of 2–3 mm. The control specimens were not shown such a pattern; the color was dark red and was uniformly distributed in the entire mass of the sample. Figures 2.3 and 2.4 show photographs of some control and osmotically treated specimen. The formation of the light-colored layers in the surface of the specimens is indicative of dead cells. The process temperature was 20°C, and processing time of the observations was 30 and 90 mins. Therefore, it is reasonable to assume that the first layers of cell in a tissue immersed in a hypertonic sucrose solution die because of osmotic shock. Given that the existence of membrane selectivity could prevent an extensive solid uptake from the medium and extensive leaching of valuable nutrients, it seems important to explore further the loss of cell viability in tissues subjected to osmotic treatments as a means to understand and, if possible, control the phenomenon.

CONCLUSIONS

(1) The intercellular space accessibility was found to differ between the two varieties examined, and it was helpful in explaining the hysteresis that the water loss curves of the Mutsu variety were shown in comparison with Kim during the osmotic processing.

(2) The state of the membrane selectivity, as determined by measurement of the EC, is another significant raw material property. The higher the EC the higher the solid gain will be. This could be explained by considering that the higher EC reflects higher age of the cells, thus being more susceptible to cell death.

(3) Our results from the actual experimental samples show that during osmotic processing the first layers of cells near the tissue-medium interface die because of severe osmotic shock. We intend to continue exploring the loss of cell viability during processing because it seems to be significant for the control of the solid gain.

(4) Furthermore, it seems possible to use the parameters explored in the present study as additional criteria among others, in the selection of suitable apple varieties.

ACKNOWLEDGEMENTS

This work was supported by the Swedish Council for Forestry and Agricultural Research (SJFR). Author K.-M. Lee thanks the Swedish Foundation for International Co-operation in Research and Higher Education (STINT) for financing his masters studies.

REFERENCES

Le Maguer, M. (1997). Mass transfer modelling in structured foods. In *Food Engineering 2000*, Fito, P., Ortega-Rodriguez, E., and Barbosa-Cánovas, G., Eds., Chapman and Hall, pp. 253–269.

Marcotte, M. and Le Maguer, M. (1992). Mass transfer in cellular tissues. Part II. Computer simulations vs experimental data. *J. Food Eng.* 17:177–199.

Mavroudis, N. E., Gekas, V., and Sjöholm, I. (1998a). Osmotic dehydration of apples—effects of agitation and raw material characteristics. *J. Food Eng.* 35:191–209.

Mavroudis N. E., Gekas V., and Sjöholm I. (1998b). Osmotic dehydration of apples: Shrinkage phenomena and the significance of the initial structure on mass transfer rates. *J. Food Eng.* 38:101–123.

Mavroudis, N. E., Wadsö, L., Gekas, V., and Sjöholm, I. (1998c). Shrinkage, microscopic studies and kinetics of apple fruit tissue during osmotic dehydration. In *Proceedings of the 11th International Drying Symposium, IDS98*. Thessaloniki-Halkidiki, Greece, August 19–22, 1998; Vol. A, pp. 844–851.

Raoult-Wack, A. L. (1994). Recent advances in the osmotic dehydration of foods. *Trends Food Sci. Technol.* 5:255–260.

CHAPTER 3

Stability of Lycopene in Tomato Dehydration

J. X. SHI
M. LE MAGUER

INTRODUCTION

TOMATO is an important agricultural commodity worldwide. More than 80% of processing tomatoes produced are consumed in the form of processed products such as tomato juice, paste, puree, catsup, sauce, and salsa (Gould, 1992). Tomatoes and tomato products are the major sources of lycopene compounds and are considered to be important contributors of carotenoids to the human diet. Consumers, researchers, and the food industry have dramatically increased their interest and awareness of the health benefits of lycopene from tomatoes. Lycopene is able to function as an antioxidant and exhibits a physical quenching rate constant with singlet oxygen *in vitro*. The quenching constant of lycopene was found to be more than double that of β-carotene and 10 times more than that of α-tocopherol, which makes its presence in the diet of considerable interest (Di Mascio et al., 1991; Conn et al., 1991; Devasagayam et al., 1992; Ribaya-Mercado et al., 1995). Increasing clinical evidence supports the role of lycopene as an important micronutrient, because it appears to provide protection against prostate cancer, lung cancer, and a broad range of other epithelial cancers (Micozzi et al., 1990; Olson, 1986; Levy et al., 1995).

The amount of lycopene in fresh tomato fruits depends on variety, maturity, and environmental conditions. Normally, tomatoes contain about 3–10 mg lycopene per 100 g raw material (Hart and Scott, 1995; Tonucci et al., 1995; Liu and Luh, 1977). McCallum (1955) studied the distribution of ly-

The work of Dr. John Shi, Dr. Marc Le Maguer, for the Department of Agriculture and Agri-Food, Government of Canada, © Minister of Public Works and Government Services Canada, 1999.

copene and other carotenoids in the tomato and found that the outer pericarp was highest in lycopene and total carotenoids, and the locular was the highest in carotene. According to Al-Wandawi et al. (1985), tomato skins contain 12 mg/100 g (wet basis) lycopene, whereas whole mature tomato contains only 3.4 mg/100 g (wet basis) lycopene. The concentration of lycopene in tomato skins is three times higher than in whole mature tomatoes.

In tomato fruits, more than 21 pigments in the carotenoid class have been identified and quantified. Lycopene is the principal carotenoid in tomatoes, with lower amounts of α-carotene, β-carotene, γ-carotene, ξ-carotene, phytoene, phytofluene, neurosporene, lutein, etc. (Gould, 1992). The chemical structures of lycopene is presented in Figure 3.1. Lycopene belongs to the subgroup of carotenes consisting only of hydrogen and carbon atoms. The chemical formulation of lycopene is $C_{40}H_{56}$. In its molecular structure, lycopene is a polyene hydrocarbon, an acyclic carotenoid having 13 double bonds, of which 11 are conjugated double bonds arranged linearly in the *trans*-form and seven of which can isomerize from the *trans*-form to the *cis*-form or vice versa under the influence of heat or certain catalysts. Lycopene has no provitamin A activity because of the lack of β-ionone ring structure of β-carotene. In nature, lycopene is almost exclusively found in the *trans*-form. Stereoisomeric forms of lycopene were described with special reference to the properties of light absorption in relation to their molecular structures. Color and antioxidant activities of lycopene are a consequence of their unique structure, an extended system of conjugated double bonds. Lycopene, by virtue of its acyclic structure and extreme hydrophobicity, will exhibit many unique and distinguishing biological features in mammalian system.

Lycopene, as a conjugated polyene, may be expected to undergo at least two changes during tomato processing, i.e., isomerization and oxidation. Lycopene isomerizations have been shown to take place both in isolated forms and in tomato products. Lycopene isomerization (from *trans* to *cis*) can take place in processing such as heating and drying. On the other hand, *cis*- and *trans*-isomers re-isomerization is another reaction during the storage of the tomato product. *cis*-Isomers are in the unstable, low-energy potential state, whereas *trans*-isomers are in the stable ground state. Although general degradation of lycopene occurs, the final products obtained are the results of direct oxidative scission at the sites of double bonds in the molecule.

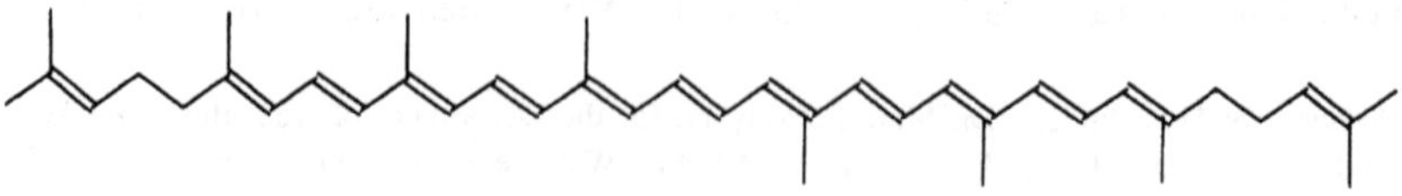

Figure 3.1 The chemical structure of lycopene molecule.

In tomato processing, tomatoes are washed, sorted, and sliced. For dried tomato slices and powder, tomatoes undergo a dehydration process. Either thermal and mechanical effects are often involved in the processes, which affects the quality of tomato products. Because color is an important quality factor for processing tomatoes, color measurements have been a convenient means of evaluating the quality of tomato products. Deep-red tomato fruits, which contain high concentrations of lycopene, are processed into products with dark-red color. The changes of lycopene content and the distribution of *trans-* and *cis-*isomers will result in change of biological property (Zechmeister, 1962).

Determination of the degree of lycopene isomerization would gain a better insight on the potential nutritional quality of the processed tomato products. In processed tomato products, oxidation is a complex process and depends on many factors, such as processing conditions, moisture, temperature, and the presence of pro- or autoxidants and lipids. The amount of sugar, acids (pH), and amino acids also affect the color of processed tomato products by causing the formulation of brown pigments (Gould, 1992). Exposure to air at high temperatures during processing of tomato products cause the naturally occurring *trans-*lycopene to be isomerized and oxidized, resulting in a loss of red color. Studies on the effect of processing conditions on qualitative and quantitative changes in lycopene degradation during tomato processing are few. This study was designed to define the effect of dehydration treatments on the lycopene content and the distribution of the *trans-* and *cis-*isomers in the tomato products in the different dehydration methods and steps for maximum retention of lycopene bioactivity in tomato products.

MATERIALS AND METHODS

MATERIAL

Mature and firm tomatoes, *Lycopersicum esculentum,* var. Heinz 9478, were obtained from the Greenhouse and Processing Crop Research Center, Agriculture and Agri-Food Canada, Harrow, Ontario, and stored at 5°C before use. Damaged and overmature fruits were discarded. The all *trans-*lycopene standard were purchased from Sigma Chemical Co. (St. Louis, MO). All reagents were of high-performance liquid chromatography (HPLC) grade.

METHODS

Skin Treatment

Tomatoes were perforated with a set of fine needles to create pin holes on the tomato surface. The pin hole density was 20 holes/cm^2 (Shi et al., 1998).

Dehydration Treatments

Two kinds of samples, intermediate moisture (IM) and dried samples, were prepared. After skin treatments, the dehydration treatments of the ripe tomatoes were designed as following three methods: (a) by conventional air drying at 95°C for 6–10 h; (b) by vacuum drying at 55°C for 4–8 h; (c) first by an osmotic treatment at 25°C in 65°Brix sucrose solution for 4 h, followed by vacuum drying at 55°C for 4–8 h. Tomatoes also underwent air drying at 90, 110, 120, and 150°C for 1–6 h, respectively. The final moisture content of IM and dried samples were 50–55% and 3–4%, respectively.

Sample Preparation for HPLC Analysis

Five tomatoes were selected and blended into puree in a Waring Blender for 3 min. The puree was homogenized with Poytron (PT2000, Kinematica Ag, Littau, Switzerland). Ten g of reconstituted tomato puree (8–9°Brix) were precisely weighted and transferred into125-ml flasks. Flasks were wrapped with aluminum foil to exclude light. One hundred ml of hexane-acetone-ethanol solution (2:1:1 v/v/v) was added into flask to solubilize lycopene, which was agitated for 10 min on a wrist action shaker until the lycopene is completely extracted. Furthermore, 15 ml of water was added followed by another 5 min on the shaker. The solution was separated into 65 ml polar and 55 ml non-polar layers, which was then followed by vacuum filtration through 0.22-μm filter paper. The upper hexane layer was collected for HPLC analysis. The entire procedure was performed in dim light. All extracts were stored in the freezer at -20°C before HPLC analysis.

Instrument and Chromatography

The content of lycopene in fresh and processed tomato samples was determined by HPLC, with an analytical 3-μm polymeric C_{30} column (C_{30} isocratic separation 4.6 mm i.d. × 250 mm, S-5) (YMC, Inc. Wilmington, NC). A mobile phase of methanol:methyl-butyl ether (MTBE) (62:38 v/v) was used at a flow of 1 ml/min. Each sample was analyzed in triplicate. Analyses were performed under dim light to prevent sample degradation by photooxidation.

Color Parameter Measurement

The color parameter measurement of fresh and dehydrated tomatoes was determined by direct reading with a Minolta Chromameter (CR200, Minolta, Japan). The instrument had an area of view of 25.4 mm and was used with a D65 illuminant as a reference at an observation angle of 10°. The "CIF lab" color parameters were calculated from the reflectance data: luminance ($L*$), red saturation index ($a*$) and yellow saturation index ($b*$).

RESULTS AND DISCUSSIONS

EFFECT OF DEHYDRATION TECHNIQUES ON LYCOPENE RETENTION

Total lycopene content in the fresh and dehydrated tomatoes is shown in Table 3.1. The dehydration of tomato slices is typically conducted at high temperatures over an extended period under vacuum. The general tendency of lycopene retention in samples slightly decreased during the dehydration processes. During osmotic dehydration, lycopene content remained essentially constant. After osmotic-vacuum drying, total lycopene retention in tomatoes was greater than those using by vacuum drying. A probable explanation is that the sugar solution keeps oxygen from the tomatoes and reduces the oxidation of lycopene in the tomato tissue matrix at low operating temperature. Conventional air drying decreases lycopene retention greatly in tomato samples. This was attributed to the influence of heat and oxygen. Heat treatment disintegrated tomato tissue and increased exposure to oxygen and light, which resulted in the destruction of lycopene.

INFLUENCE OF DEHYDRATION ON LYCOPENE ISOMERIZATION

The distribution of isomers in the different dehydrated tomato samples are shown in Table 3.1. In the fresh tomato samples, lycopene content is 755 μg/g on a dry weight basis. *cis*-Isomers were not detected in the fresh tomato samples. Lycopene occurs in nature primarily in the more stable *trans*-configuration (Zechmeister, 1962; Chandler and Schwartz, 1987; Rodriguez-Amaya and Tavares, 1992). The *cis*-isomers appeared in processed tomato samples (Figure 3.2). A significant increase in the *cis* isomers with simultaneous decrease in the *trans*-isomers was observed in the dehydrated tomato samples through different dehydration methods.

TABLE 3.1. Total Lycopene and *cis*-Isomer Content in the Dehydrated Tomato Samples (Shi et al., 1999).

Sample	Total Lycopene (μg/g Dry Basis)	Lycopene Loss (%)	*trans*-Isomers (%)	*cis*-Isomer (%)
Fresh tomato	755[a]	0	100	0
Osmotic treatment	755[a]	0	100	0
Osmo-vac dried	737[b]	2.4	93.5	6.5
Vac-dried	731[c]	3.2	89.9	10.1
Air-dried	726[d]	3.9	84.4	16.6

Data are presented as means of triplicate determinations.
Means in a column not sharing common superscript (a–d) are significantly different ($P < 0.01$).

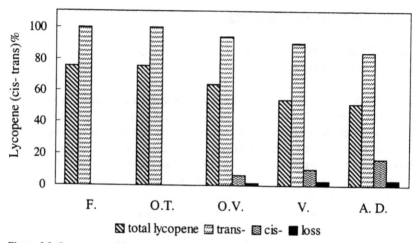

Figure 3.2 Comparison of lycopene degradation in the different dehydration processes (F—fresh tomato, O.T.—osmotic treatment, O.V.—osmotic-vacuum drying, V—vacuum drying, A.D.—air drying) (Shi et al., 1999).

Degradation of lycopene not only affects the attractive color of final products but also their nutritive value for health benefit. The main cause of lycopene biodegradation in tomato dehydration is isomerization and oxidation. It is widely presumed that lycopene in general undergoes isomerization with thermal processing. This isomerization resulted in conversion of *trans*- isomers to *cis*-isomers. It was observed that fewer *cis*-isomers were present in osmotically dehydrated tomatoes compared with those directly air dried and vacuum dried. Tomatoes after osmotic treatment showed very little isomerization. Other dehydration methods, especially in conventional air drying, produced greater isomerization. The highest amount of *cis*-isomers was found in air-dried tomato samples. This supports the assumption of Miers et al. (1958) that the amount of *cis*-isomers exceeds those present in the initial tomato material and osmotically treated tomato samples when dehydrated by conventional methods. The *cis*-isomers were formed in processed tomato samples and increased with temperature and time during dehydration. An increase in *cis*-isomers indicates a loss of biopotency of lycopene. Each sample dehydrated by different methods had a negative factor favoring isomerization and/or oxidation of the lycopene, e.g., oxygen permeability, light exposure, and perhaps presence of some metals in the processing system. A large loss of lycopene during processing would result from a longer and more drastic procedure, particularly in the thermal dehydration steps. Dehydration of tomatoes at mild temperature does not usually cause significant loss in total lycopene content (Nguyen and Schwartz, 1998), but the conversion of *trans*- to *cis*-isomers always occurred in the dehydrated products. In the osmotic treatment, the predominating mechanism may be isomerization of lycopene. Because the total lycopene content remained almost constant, only the distribution of *trans-* and *cis*-isomers was changed. In air drying, isomerization and oxidation (autoxidation) were two strong factors that simultaneously affected the total lycopene content, distribution of *trans-* and *cis*-isomers, and biological potency.

It should be pointed out that lycopene is a polyene component having 13 double bonds, of which 11 are conjugated double bonds and 7 of which can isomerize from the *trans*-form to the *cis*-form or vice versa under the influence of heat, light, mechanical action, and other factors. The changes in lycopene content and the distribution of *trans-* and *cis*-isomers will result in a reduction in biological potency, when tomato-based products are subjected to processing (Zechmeister, 1962; Khachik et al., 1992; Emenhiser et al., 1995; Wilberg and Rodriguez-Amaya, 1995; Stahl and Sies, 1996). Osmotic solution (sugar) remaining on the surface layer of tomato prevents oxygen from penetrating and oxidizing lycopene. Osmotic treatment could reduce lycopene losses in comparison with other dehydration methods. These results will be useful to develop new dehydration techniques and improve product quality.

EFFECT OF TEMPERATURE ON LYCOPENE DEGRADATION

The effects of temperature on total lycopene and *cis*-isomer content in tomato puree during dehydration are shown in Figures 3.3 and 3.4. Increasing the temperature of the heat treatment from 90 to 150°C caused a greater decrease in total lycopene. It was also observed that most of the changes in the concentration of total lycopene and *cis*-isomer occurred within the first hour of heat treatment. After 2 h of the heat treatment, the rate of degradation decreased. The temperature increase from 90°C to 150°C caused a 35% de-

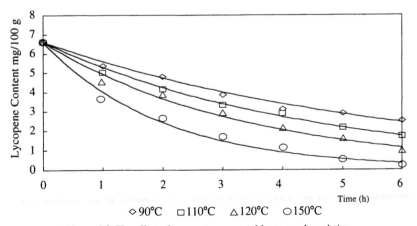

Figure 3.3 The effect of temperature on total lycopene degradation.

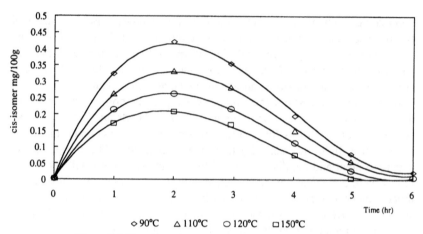

Figure 3.4 The effect of temperature on *cis*-isomer degradation.

crease in total lycopene. The increase in temperature causes the degradation and isomerization of lycopene. The greater percentage of lycopene loss compared with the gain in *cis*-isomer suggest that oxidation of lycopene was the main mechanism for the lycopene loss during heat treatment. Fewer *cis*-isomers were present in the low-temperature treatment. The *cis*-isomers were formed in processed tomato samples and increased with temperature and time during the first 1- to 2-h treatment. When tomato-based products are subjected to processing, the changes in lycopene content and the distribution of *trans* form to *cis*-isomers may result in a change in bioactivity potency (Zechmeister 1962; Khachik et al., 1992; Emenhiser et al., 1995; Wilberg and Rodriguez-Amaya, 1995; Stahl and Sies, 1996).

LYCOPENE DEGRADATION AND COLOR CHANGES OF TOMATO PRODUCTS

A number of publications have reported the tendency of lycopene compounds to isomerize from one form to another with accompanying color changes (Wong and Bohart, 1957; Miers et al., 1958). Lycopene is located in chromoplasts dispersed throughout the tomato fruits. Lycopene appears as solid microcrystals so that the light reflected from them gives the tomato its typical bright-red color. When lycopene is dissolved in lipids or other solvents, its color is yellow or dark orange, but not red. It seems possible that the naturally occurring *trans*-lycopene isomerizes to the less red partly *cis*-isomers with a corresponding change in absorption spectra during processing of tomato products (Miers et al., 1958). Color evaluation of whole fresh tomatoes have traditionally been presented as Hunter L^*, a^*, b^* values. Results of color parameters L^*, a^*, and b^* together with the ratio a^*/b^* and the overall color difference (ΔE) of the dehydrated tomato products are presented in Table 3.2. Tomatoes with osmotic treatment had more red color than those treated by air drying and vacuum drying, which indicated there was more lycopene in the samples. For the overall color difference ΔE,

$$\Delta E = \sqrt{(\Delta L^*)^2 + (\Delta a^*)^2 + (\Delta b^*)^2}$$

Wiese and Dalmasso (1994) reported an increase in the hue angle of tomato juice after processing and storage, indicating loss of red color. Color retention in tomato products is better at lower temperatures (Sherkat and Luh, 1976; Villari et al., 1994). The main cause of lycopene degradation in foods is oxidation. In processed tomato products, oxidation is a complex process and depends on many factors, such as processing conditions, moisture, oxygen, temperature, light, and the presence of pro- or autoxidants and lipids. The large surface exposed to air and metal enhanced oxidation of pigments of tomato products (Miers et al., 1958). The amount of sugar, acids (pH), and amino acids, as well as time of processing also affected the color of processed tomato products by causing the formation of brown pigments (Gould, 1992). From Table 3.2, a slightly better color can be observed in the samples dehydrated at low temperatures. The color differences between the samples were not readily discernible by visual evaluation. There was no significant difference between Hunter color value a^* of different dehydrated tomatoes. This was attributed to the lycopene crystals formation in the tomato tissue matrix after heating in the dehydration processes. On heating, the spectrum did not change greatly. But there was a significant difference ($P = 0.01$) in the ratio of *trans*- to *cis*-isomers. The color measurement

TABLE 3.2. Color Values of Dehydrated Tomato Samples (Shi et al., 1999).

Samples and Dehydration Condition	Color Parameters					
	L^*	b^*	b^*	a^*/b^*	ΔE	L^*b^*/a^*
Fresh material	38.4[a]	37.7[a]	16.1[a]	2.3[a]	56.2[a]	16.4[a]
IM tomatoes						
Osmotic dehydration at 25°C	38.4[a]	37.7[a]	16.1[a]	2.3[a]	56.2[a]	16.4[a]
Osmo-vac drying at 55°C	36.7[a]	35.2[a]	16.9[a]	2.1[a]	53.6[a]	17.2[a]
Vacuum drying at 55°C	34.2[a]	34.3[a]	17.4[a]	1.9[a]	51.5[b]	17.4[b]
Air drying at 95°C	29.9[b]	33.2[b]	19.9[b]	1.7[b]	48.9[c]	18.1[c]
Dehydrated tomatoes						
Osmo-vac drying at 55°C	31.4[c]	36.4[c]	18.3[b]	1.7[b]	48.1[b]	18.3[c]
Vacuum drying at 55°C	28.3[d]	25.6[d]	18.2[b]	1.4[b]	42.3[c]	20.1[d]
Air drying at 90°C	25.6[e]	23.2[e]	18.9[b]	1.2[b]	39.4[d]	20.9[e]

Data are presented as means of triplicate determinations.
Means in a column not sharing common superscript (a–e) are significantly different ($P < 0.01$).

did not show the relative composition of *trans-* and *cis*-isomers. An increase in *cis*-isomers would indicate a change in lycopene bioactivity but would not show up as a significant difference in color. Lycopene content and the ratio of *trans*- to *cis*-isomers may have caused the $a*/b*$ value to stay at a higher level (Wong and Bohart, 1957). The color quality, $a*/b*$ values, remained essentially unchanged during the osmotic treatment, but there were lower values of $a*/b*$ in the conventional air-dried sample. Product color showed progressive deterioration of overall color quality (ΔE) in conventional air drying.

Yeatman (1969) indicated that the value $b*L*/a*$ provided a high linear correlation with visual color scores of processed tomato products. The overall mean $b*L*/a*$ value for the osmotic treatment, osmo-vac drying, were 16.42 and 18.34, respectively. The average color reflectance reading for osmotically dehydrated fruits had color characteristics close to those of the fresh material. The $L*$ and $a*$ values decreased in the other dehydration treatment. A comparison of lycopene degradation tendency and color parameters in the different dehydrated tomato products show the change tendencies are not parallel. Tomatoes with osmotic treatment had more red color than those treated by air drying and vacuum drying, which indicated there was more lycopene in the samples. Osmotically dehydrated tomatoes appeared to be promising through this new processing technique to keep the fresh natural reddish color.

CONCLUSIONS

Conservation of lycopene during tomato processing of tomato products is of commercial significance. Degradation of lycopene not only affects the attractive color of tomato products but also their nutritive value and flavor. Four dehydration methods produced slight differences in the total lycopene content but resulted in quite different distribution of the isomer composition. Osmotic treatments retained more total lycopene and induced only slight changes in the distribution of *trans*- and *cis*-isomers. The osmotic-vacuum treatment had less effect on lycopene loss and isomerization than vacuum drying and conventional air drying. Heat treatment under atmospheric conditions in the dehydration processes accounts for the lycopene degradation through isomerization and oxidation. In the osmotic treatment, the predominating mechanism of change may be isomerization of lycopene. Because the total lycopene content remained essentially constant, only the distribution of *trans*- and *cis*-isomers was changed. In the air drying, isomerization and oxidation (autoxidation) were two factors that simultaneously affected the decrease of total lycopene content, distribution of *trans*- and *cis*-isomers, and biological potency. Osmotic treatment could reduce lycopene losses in comparison with other dehydration methods. These results will be useful to develop new dehydration technologies and improve product quality.

ACKNOWLEDGEMENT

This study received support from Southern Crop Protection and Food Research Center, Agriculture and Agri-Food Canada (AAFC).

REFERENCES

Al-Wandawi, H., Abdul-Rahman, M., and Al-Shaikhly, K. 1985. Tomato processing waste as essential raw materials source. *J. Agric. Food Chem.* 33:804–807.

Chandler, L. A. and Schwartz, S. J. 1987. HPLC separation of cis-trans carotene isomers in fresh and processed fruits and vegetables. *J. Food Sci.* 52(3):669–672.

Conn, P. F., Schalch, W., and Truscott, T. G. 1991. The singlet oxygen andcarotenoid interaction. *J. Photochem. Photobiol. B. Biol.* 11:41–47.

Devasagayam, T. P. A., Werner, T., Ippendorf, H., Martin, H. D., and Sies, H. 1992. Synthetic carotenoids, novel polyene polyketones and new capsorubin isomers as efficient quenchers of singlet molecular oxygen. *Photochem. Photobiol.* 55:511–514.

Di Mascio, P., Murphy, M. C., and Sies, H. 1991. Antioxidant defense systems, the role of caritenoid, tocopherol and thiols. *Am. J. Clin. Nutr., Suppl.* 53:194–200.

Emenhiser, C., Sander, L. C., and Schwartz, S. J. 1995. Capability of a polymeric C_{30} stationary phase to resolve cis-trans carotenoids in reversed phase liquid chromatograph. *J. Chromatogr. A* 707:205–216.

Gould, W. A. 1992. *Tomato Production, Processing and Technology.* Baltimore: CTI Publisher.

Hart, D. J. and Scott, K. J. 1995. Development and evaluation of an HPLC method for the analysis of carotenoids in foods, and the measurement of the carotenoid content of vegetables and fruits commonly consumed in the UK. *Food Chem.* 54:101–111.

Khachik, F., Beecher, G. R., Goli, N. B., Luby, W. R., and Smith, J. C. 1992. Separation and identification of carotenoids and their oxidation products in the extracts of human plasma. *Anal. Chem.* 64:2111–2122.

Levy, J., Bisin, E., Feldman, B., Giat, Y., Miinster, A., Danilenko, M., and Sharoni, Y. 1995. Lycopene is a more potent inhibitor of human cancer cell proliferation then either a-cartotene or b-carotene. *Nutr. Cancer* 24(3):257–266.

Liu, Y. K. and Luh, B. S. 1977. Effect of harvest maturity on carotenoids in pastes made from VF-145-7879 tomato. *J. Food Sci.* 42:216–220.

McCallum, J. P. 1955. Distribution of carotenoids in the tomato. *Food Res.* 20:55–59.

Micozzi, M. S., Beecher, G. R., Taylor, P. R., and Khachik, F. 1990. Carotenoid analyses of selected raw and cooked foods associated with a lower risk for cancer. *J. Natl. Cancer Inst.* 82:282–288.

Miers, J., Wong, F., Harris, J., and Dietrich, W. C. 1958. Factors affecting storage stability of spray-dried tomato powder. *Food Technol.* 10:542–548.

Nguyen, M. and Schwartz, S. J. 1998. Lycopene stability during food processing. *Proc Soc. Exp. Biol. Med.* 218:101–105.

Olson, J. 1986. Carotenoid, vitamin A and cancer. *J. Nutr.* 116:1127–1130.

Ribaya-Mercado, J. D., Garmyn, M., Gilchrest, B. A., and Russell, R. M. 1995. Skin lycopene is destroyed preferentially over β-carotene during ultraviolet irradiation in humans. *J. Nutr.* 125:1854–1859.

Rodriguez-Amaya, D. B. and Tavares, C. A. 1992. Importance of *cis*-isomer separation in determining provitamin A in tomato and tomato products. *Food Chem.* 45:297–302.

Sherkat, F. and Luh, B. S. 1976. Quality factors of tomato paste made at several break temperatures. *J. Agr. Food Chem.* 24:1155–1158.

Shi, J., Le Maguer, M., Wang, S., and Liptay, A. 1998. Application of osmotic treatment in tomato processing—effect of skin treatments on mass transfer in osmotic dehydration of tomatoes. *Food Res. Int.* 30(9):669–674.

Shi, J., Le Maguer, M., Kakuda, Y., Liptay, A., and Nickamp, F. 1999. Lycopene degeneration and isomerization in tomato degradation. *Food Research International* 32:15–21.

Stahl, W. and Sies, H. 1996. Perspectives in biochemistry and biophysics. *Arch. Biochem. Biophys.* 336(1):1–9.

Tonucci, L. H., Holden, J. M., Beecher, G. R., Khachik, F., Davis, C., and Mulokozi, G. 1995. Carotenoid content of thermally processed tomato-based food products. *J. Agric. Food Chem.* 43:579–586.

Villari, P., Costabile, P., Fasanaro, G., Desio, F., Laratta, B., Pirone, G., and Castaldo, D. 1994. Quality loss of double concentrated tomato paste: Evolution of the microbial flora and main analytical parameters during storage at different temperature. *J. Food Processing Preservation,* 18:369–387.

Wiese, K. L. and Dalmasso, J. P. 1994. Relationship of color, viscosity, organic acid profiles and ascorbic acid content to addition of organic acid and salt in tomato juice. *J. Food Quality* 17:273–284.

Wilberg, V. C. and Rodriguez-Amaya, B. D. 1995. HPLC quantitation of major carotenoids of fresh and processed guava, mango and papaya. *Lebensmittel-Wissenschaft und Technologie* 28:474–480.

Wong, F. F. and Bohart, G. S. 1957. Observation on the color of vacuum-dried tomato juice powder during storage. *Food Technol.* 5:293–296.

Yeatman, J. N. 1969. Tomato products: read tomato red? *Food Technol.* 23:618–627.

Zechmeister, L. 1962. *Cis-Trans Isomeric Carotenoids, Vitamin A and Arylpolyenes.* New York: Academic Press.

CHAPTER 4

Reasons and Possibilities to Control Solids Uptake during Osmotic Treatment of Fruits and Vegetables

H. N. LAZARIDES

INTRODUCTION

OSMOTIC preconcentration (dehydration) is the partial removal of water by direct contact of a product with a hypertonic medium, i.e., a high-concentration sugar or salt solution for fruits and vegetables, respectively.

During osmotic processing, two major countercurrent flows take place simultaneously. Under the water and the osmotic solute activity, gradients across the product-medium interface, and water flows from the product into the osmotic solution, whereas osmotic solute is transferred from the solution into the product (Figure 4.1).

A third transfer process, leaching of product solutes (sugars, acids, minerals, and vitamins) into the solution, although recognized as affecting the organoleptic and nutritional characteristics of the product (Dixon and Jen, 1977), it is considered quantitatively negligible.

As the demand for competitive (high-quality and low-cost) products increases, there is a progressively greater pressure to minimize the negative impact and cost of processing, as well as the cost of shipping, handling, and storing of agricultural products.

Osmotic preconcentration is an effective way to reduce water content with minimal damage on fresh product quality.

This is largely due to the use of a mild product treatment at relatively low process temperatures (up to 50°C); such temperatures do not affect the semipermeable characteristic of cell membranes, which is an essential requirement for maintaining the osmotic phenomenon.

Moderate temperatures also favor color and flavor retention resulting in products with superior organoleptic and nutritional characteristics (Ponting,

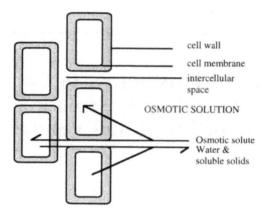

cell wall

cell membrane

intercellular
space

OSMOTIC SOLUTION

Osmotic solute
Water &
soluble solids

Figure 4.1 Mass exchange between natural tissue and osmotic solution during osmotic processing (adapted from Lazarides, 1994).

1973). Sugars are known to prevent loss of volatile flavor components during vacuum drying (Wienjes, 1968). Thus, limited sugar uptake (during osmotic processing) favors aroma retention in osmotic-vacuum dehydrated fruits.

Because of constant product immersion in the osmotic medium, the plant tissue is not exposed to oxygen; therefore, there is no need to use antioxidants (i.e., sulfur dioxide) for protection against oxidative and enzymatic discoloration (Ponting et al., 1966; Ponting, 1973; Dixon et al., 1976).

Because water is removed in liquid (and not in vapor) form, in overall energy requirements, osmotic dehydration is superior to other dehydration processes aiming at product stabilization. It has been found that osmotic dehydration with syrup reconcentration demands two to three times less energy (per unit) than convection drying (Lenart and Lewicki, 1988).

If semidehydration is followed by freezing, energy demands for water freezing are cut in half, and packaging, transportation, and handling costs are decreased substantially (Lazarides, 1995).

The above advantages make osmotic preconcentration an attractive process step in several alternative processing schemes.

OSMOTIC PROCESS APPLICATIONS

Based on the desired final product characteristics, the osmotic process may emphasize water removal, solute uptake, or a certain balance between water removal and specific functional solute uptake. Under these alternative process approaches, the entire range of osmotic process applications can be classified in the following categories:

(1) Partial water removal (solids concentration) followed by:
  • pasteurization (and cold storage)

- freezing
- complimentary dehydration (air, vacuum, freeze, or microwave)

(2) Solute impregnation with:
- sugars (candying)
- salts (salting)

(3) Product formulation aiming at:
- better organoleptic quality
- texturization (enhancing textural characteristics by incorporation of selected texturizing agents, i.e., calcium ions, enzymes, etc.)
- better nutrition (nutritional supplements)
- microbial stability (antimicrobials)
- combinations of the above aims

(4) Combinations of the above three categories in successive processing steps

IMPACT OF SOLUTE UPTAKE

Solute uptake during osmotic dehydration modifies the composition and taste of the final product. This so-called "candying" (or "salting") effect is sometimes desirable, because it tends to improve the taste and acceptability of the final product.

In many cases, however, extensive solute uptake is undesirable, because of its negative impact on taste and the nutritional profile of the product, which can no longer be marketed as "natural." Leaching of natural acids out of osmotically dehydrated fruits also affects taste, because it changes the natural sugar to acid ratio.

In every case, solute uptake results in the development of a concentrated solids layer under the surface of the fruit, upsetting the osmotic pressure gradient across the fruit-medium interface and decreasing the driving force for water flow (Hawkes and Flink, 1978).

Besides its negative impact on the rate of water loss during osmotic preconcentration, solute uptake "blocks" the surface layers of the product, posing an additional resistance to mass exchange and lowering the rates of complimentary (vacuum, convection, or freeze) dehydration (Lenart and Grodecka, 1989).

On the other hand, partial dehydration and solute uptake protect the product against structural collapse during complimentary dehydration and against structural disorganization (collapse) and extensive exudation (dripping) on freeze/thawing (Lazarides and Mavroudis, 1995). Structural stability could be further enhanced by incorporation of cryoprotectants (Martinez-Monzo et al., 1998).

Solute uptake also affects product rehydration (hygroscopicity). Rehydration of osmotically dried fruit is lower than the untreated in both rate and extent (Ponting et al., 1966; Lerici et al., 1988; Lenart, 1991; Lazarides et al.,

1995b). This is due to the lower rehydration of sugar in the product, compared with the natural tissue itself. It has been shown that the longer the osmosis time, the lower the rehydration rate and extent of osmotic-convection dried carrots (Lenart, 1991). Water vapor adsorption followed a similar pattern.

The above discussion underlines the importance of the rate and extent of solute uptake in managing the rate of water removal and in achieving our process objectives (i.e., reach the desired quality characteristics, the stability and the functional properties of the final product).

In short, the following effects can be recognized on the product side and on the process side:

a. Product side: sensorial effect, structural effect, nutritional effect, natural composition profile, chemical, structural, and microbial stability, and functional properties
b. Process side: control water removal rates and affect complimentary processes (i.e., air dehydration)

OPTIMUM LEVELS OF SOLUTE UPTAKE

Optimum level of solute uptake strongly depends on process objectives and desired final product properties. Relative extent of solute uptake is usually expressed in terms of so-called *"dehydration efficiency index,"* which is the ratio of water loss/solid gain (WL/SG).

Whenever our goal is monodimentional, i.e., either dehydration or impregnation, we need to establish conditions promoting water removal or solute gain, respectively.

Therefore, if we are after fast water removal, we need to reach a high WL/SG ratio, keeping SG as low as possible, always below a maximum value that is set by the desired properties of the final product.

In case of impregnation, we proceed in the opposite direction, because we seek a low WL/SG ratio, keeping WL as low as possible, to both minimize weight loss and induce a desired textural effect (i.e., water plasticizing).

If our process goal is polydimentional, as in the case of formulation, we need to control the WL/SG ratio at a level covering our formulation needs. In such cases, SG must be kept between a minimum and a maximum value, according to our specific product needs.

POSSIBILITIES TO CONTROL SOLIDS UPTAKE

We can recognize two groups of parameters affecting solute uptake: product parameters and process parameters.

a. Product parameters include:
- fresh material characteristics (porosity, structural elements), depending on: species, variety, maturity, cultivation measures, and soil-climate conditions
- size/shape (specific surface)
- pretreatment: peeling or coating, blanching, and freeze/thawing

b. Process parameters include:
- solution composition (binary, ternary, molecular size, etc.)
- solution concentration
- process temperature
- process pressure (atmospheric, vacuum)
- process time
- product/solution contacting (agitation, product/solution ratio)
- use of ultrasound (sonication)

EFFECT OF PRODUCT PARAMETERS ON SOLUTE UPTAKE

On the product side, species, variety, and maturity level all have a significant effect on natural tissue structure in cell membrane structure, protopectin to soluble pectin ratio, amount of insoluble solids, intercellular spaces, tissue compactness, entrapped air, etc. Structural differences are not only encountered among materials of different species, variety, or maturity level but even between different sections of the same piece of fruit (i.e., apples) (Vincent, 1989).

Tissue density is the most important material characteristic in thermophysical and transport properties (Rahman et al., 1996). Therefore, structural differences substantially affect diffusional mass exchange between product and osmotic medium.

Indeed, structure (especially porosity) of the raw material had a significant effect on both shrinkage phenomena and mass transfer rates (Lazarides and Mavroudis, 1996; Lazarides et al., 1997; Lazarides, 1998; Mavroudis et al., 1998a).

The geometry and size of product affect the surface/volume ratio (specific surface). High specific surface values favor solute impregnation, because solute uptake is a surface-controlled process (Lerici et al., 1985; Torreggiani, 1993). According to Lerici et al. (1985), up to a certain *A/L* (total surface/half thickness) ratio, higher specific surface samples (i.e., rings) gave higher water loss (WL) and sugar gain (SG) values than lower specific surface shapes (i.e., slice and stick). Beyond that A/L limit, higher specific surface samples (i.e., cubes) favored sugar gain at the expense of lower water loss, resulting in lower weight reduction. The lowest water loss associated with the highest A/L ratio was explained as the result of reduced water diffusivity due to high sugar uptake.

Product pretreatments and process conditions affecting the integrity of natural tissue have a severe effect on water loss and solids gain. Disruption of structural barriers seems to result in decreased tortuosity of diffusion paths favoring solids transfer (Oliveira and Silva, 1992). Blanching, freeze/thawing, sulfiding, acidification, and high process temperatures all favor solids uptake yielding lower WL/SG ratios (Lerici et al., 1988; Biswal and Le Maguer, 1989; Lazarides, 1995; Lazarides and Mavroudis, 1995).

High pressure pretreatment of raw material (up to 400 Mpa) enhanced mass transfer rates, favoring higher WL/SG ratios (Rastogi and Niranjan, 1998). Microstructural examination revealed that high pressure treatment broke down cell walls and softened the tissue.

Because water and osmotic solutes flow in opposite directions and because their flows are highly antagonistic, precoating of raw materials with edible, carbohydrate polymer coatings has been used to minimize solute uptake and free the way for faster water removal (Lewicki et al., 1984; Camirand et al., 1992; Ishikawa and Nara, 1993; Lenart and Dabrowska, 1998).

EFFECT OF PROCESS PARAMETERS ON SOLUTE UPTAKE

On the process side, the osmotic solute(s) used and especially the molecular size play a very important role.

A comparison of various osmotic solutions at a constant solids concentration showed that mixed sucrose/salt solutions gave a greater decrease in product water activity than pure sucrose solutions, although water transport rates were similar (Lenart and Flink, 1984a). This was blamed on extensive salt uptake. Spatial distribution analysis by the same workers revealed large differences between osmotic solute distribution curves for dehydration taking place in sucrose or salt solutions (Lenart and Flink, 1984b)

Early penetration studies showed that the rate of solute penetration is directly related to the solution concentration and inversely related to the size of the sugar molecule (Hughes et al., 1958). By using higher molecular weight sugars (i.e., lower dextrose equivalent [DE] corn syrup solids) it was possible to zero net solute gain (Lazarides et al., 1995a). It was even possible to have a net loss of solids (solids uptake was lower than solids leaching out).

Therefore, corn syrup solids of high DE (low MW) would be the right choice for impregnation processes.

Solution concentration, process temperature, and process duration also have a severe impact on solids uptake (Torreggiani, 1993; Lazarides, 1994; Dalla Rosa et al., 1995; Lazarides and Mavroudis, 1995; Lazarides et al., 1999).

During extended osmotic treatment, increased solute concentrations resulted in increased WL and SG rates, favoring higher WL/SG ratios (Hawkes and Flink, 1978; Lenart, 1991). Evolution of WL/SG ratios was different, however, during the early stages of osmosis. On osmotic dehydration of apples in a higher concentration sugar solution for 3 h, there was some benefit in terms of faster water loss (~30% increase); at the same time, however, there was a much greater uptake of sugar solids (~80% increase). The net result was that short-term osmosis under increased concentrations favored solute uptake, resulting in lower WL/SG ratios (Lazarides et al., 1995a).

During osmotic dehydration of potatoes, increasing process temperature up to 45°C resulted in increased WL and SG rates, in favor of higher WL/SG ratios. Higher temperatures, however, had a detrimental effect on tissue structure, resulting in a drastic increase in the rate of solute uptake, followed by a marginal increase in the rates of WL (Lazarides and Mavroudis, 1996). Therefore, use of higher process temperatures is only suggested in treatments aiming at product impregnation. High temperature limits are naturally set by the heat tolerance of the specific tissue to be processed and especially the heat stability of its cell membrane.

Vacuum osmotic processing strongly favors solute uptake (impregnation), through an effective increase of mass transfer surface, caused by replacement of gas in the pores with osmotic solution (Fito et al., 1994; Chiralt et al., 1999).

A similar effect was observed, when ultrasound was applied to the osmotic system. By using sonication in osmotic processing of apples, mass transfer rates were increased, with prevailing increase of sucrose transfer rates (Simal et al., 1998). The sonication effect on increased solute uptake was equivalent to a rise in process temperature by 30°C. In other words, it was necessary to maintain the solution temperature at 70°C under agitation, to achieve a similar solute gain to that obtained at 40°C with sonication.

Degree of agitation (agitation Re number) had a significant effect on WL, but it had no substantial effect on SG (Mavroudis et al., 1998b). WL was higher in the turbulent flow region than in the laminar flow region. Therefore, turbulent solution flow due to agitation may result in higher WL/SG ratios.

CONCLUSIONS

(1) There are many reasons to control solids uptake and many more parameters controlling it.
(2) Before we set to control solids uptake, we need to carefully define the desired final product characteristics.
(3) On the basis of process goals we need to select suitable raw material, specify and monitor appropriate process conditions.
(4) Solid-liquid processes are complex unit operations, and as such, they need pilot-scale confirmation and final "tuning."

REFERENCES

Biswal, R. N. and Le Maguer, M. 1989. Mass transfer in plant materials in aqueous solutions of ethanol and sodium chloride: Equilibrium data, *J. Food Process Eng.* 11(3):159.

Camirand, W., Krochta, J. M., Pavlath, A. E., Wong, D., and Cole, M. E. 1992. Properties of some edible carbohydrate polymer coatings for potential use in osmotic dehydration. *Carbohydrate Polymers* 17:39–49.

Chiralt, A., Fito, P., Andres, A., Barat, J., Martinez-Monzo, J., and Martinez-Navarrete, N. 1999. Vacuum impregnation: A tool in minimally processing of foods. In: Oliveira, F. A. R. and Oliveira, J. C. (Eds), *Processing Foods: Quality Optimization and Process Assessment.* Boca Raton, FL: CRC Press, pp. 341–356.

Dalla Rosa, M., Bressa, F., Mastrocola, D., and Pittia, P. 1995. Use of osmotic treatments to improve the quality of high-moisture minimally processed foods. In: A. Lenart and P. P. Lewicki (Editors). *Proceedings of the Second International Seminar on Osmotic Dehydration of Fruits and Vegetables.* Warsaw, April 18–19, 1994. pp. 88–98.

Dixon, G. M. and Jen, J. J. 1977. Changes of sugars and acids of osmovac-dried apple slices, *J. Food Sci.* 42(4):1136.

Dixon, G. M., Jen, J. J., and Paynter, V. A. 1976. Tasty apple slices result from combined osmotic-dehydration and vacuum-drying process. *Food Prod. Devel.* 10(7):60, 62, 64.

Fito, P., Andres, A., Pastor, R., and Chiralt, A. 1994. Vacuum osmotic dehydration of fruits. In: Singh, R. P. and Oliveira, F. A. R. (Eds), *Minimal Processing of Foods and Process Optimization. An Interface.* Boca Raton, FL: CRC Press, pp. 107–122.

Hawkes, J. and Flink, J. M. 1978. Osmotic concentration of fruit slices prior to freeze dehydration. *J. Food Proc. Preserv.* 2:265.

Hughes, R. E., Chichester, C. O., and Sterling, C. 1958. Penetration of maltosaccharides in processed Clingstone peaches. *Food Technol.* 12:111–115.

Ishikawa, M. and Nara, H. 1993. Osmotic dehydration of food by semi-permeable membrane coating. In: Singh, R. P. and Wirakartakusuman, M. A. (Eds). *Advances in Food Engineering.* London: CRC Press, pp. 73–77.

Lazarides, H. N. 1994. Osmotic preconcentration: Developments and prospects. In: Singh, R. P. and Oliveira, F. A. R. (Eds), *Minimal Processing of Foods and Process Optimization. An Interface.* Boca Raton, FL: CRC Press, pp. 73–85.

Lazarides, H. N. 1995. Osmotic preconcentration as a tool in freeze preservation of fruits and vegetables. In: Lenart, A. and Lewicki, P. P. (Eds). *Proceedings of the Second International Seminar on Osmotic Dehydration of Fruits and Vegetables.* Warsaw, April 18–19, 1994, pp. 88–98.

Lazarides, H. N. 1998. Mass transfer phenomena during osmotic processing of fruits and vegetables. In: Oliveira J. C. and F. A. R. Oliveira (Eds). *Proceedings of the Third Main Meeting of the Copernicus Project in "Process Optimization and Minimal Processing of Foods, Vol. 3, Drying.* Leuven, October 23–25, 1997, pp. 43–45.

Lazarides, H. N.. Fito, P., Chiralt, A., Gekas, V., and Lenart, A. 1999. Advances in osmotic dehydration. In: Oliveira, F. A. R., and Oliveira, J. C. (Eds). *Processing Foods: Quality Optimization and Process Assessment.* CRC Press, pp. 175–199.

Lazarides, H. N., Gekas, V., and Mavroudis, N. 1997. Apparent mass diffusivities in fruit and vegetable tissues undergoing osmotic processing. *J. Food Eng.* 31:315–324.

Lazarides, H. N., Katsanides, E., and Nicolaides, A. 1995a. Mass transfer kinetics during osmotic preconcentration aiming at minimal solid uptake. *J. Food Eng.* 25(2):151–166.

Lazarides, H. N. and Mavroudis, N. 1995. Freeze/thaw effect on mass transfer rates during osmotic dehydration. *J. Food Sci.* 60(4):826–829.

Lazarides, H. N. and Mavroudis, N. 1996. Kinetics of osmotic dehydration of a highly shrinking vegetable tissue in a salt-free medium. *J. Food Eng.* 30:61–74.

Lazarides, H. N., Nicolaidis, A., and Katsanidis, E. 1995b. Sorption behavior changes induced by osmotic preconcentration of apple slices in different osmotic media. *J. Food Sci.* 60(2):348–350, 359.

Lenart, A. 1991. Effect of saccharose on water sorption and rehydration of dried carrot. In: Mujumdar, A. S. and Filkova, I. (Eds). *Drying '91.* Amsterdam: Elsevier Science, p. 489.

Lenart, A. and Dabrowska, R. 1998. Influence of edible carbohydrate coatings on osmotic dehydration of apples. In: Oliveira J. C. and F. A. R. Oliveira (Eds). *Proceedings of the Third Main Meeting of the Copernicus Project in "Process Optimization and Minimal Processing of Foods, Vol. 3, Drying.* Leuven, October 23–25, 1997, pp. 33–37.

Lenart, A. and Flink, J. M. 1984a. Osmotic concentration of potato. I. Criteria for the end-point of the osmosis process. *J. Food Technol.* 19(1):45.

Lenart, A. and Flink, J. M. 1984b. Osmotic concentration of potato. II. Spatial distribution of the osmotic effect. *J. Food Technol.* 19:65.

Lenart, A. and Grodecka, E. 1989. Influence of the kind of osmotic substance on the kinetics of convection drying of apples and carrots, Ann. Warsaw Agricult. Univ.-SGGW-AR. *Food Technol. Nutr.* 18:27.

Lenart, A. and Lewicki, P. P. 1988. Energy consumption during osmotic and convective drying of plant tissue. *Acta Alimentaria Polonica.* XIV(1):65.

Lerici, C. R., Mastrocola, D., Sensidoni, A., and Dalla Rosa, M. 1988. Osmotic concentration in food processing. In Bruin, S., (ed). *Proceedings of the International Symposium on Preconcentration and Drying of Foods,* Eindhoven, The Netherlands, Nov. 5–6, 1987, Elsevier, p. 123.

Lerici, C. R., Pinnavaia, G., Dalla Rosa, M., and Bartolucci, I. 1985. Osmotic dehydration of fruit: Influence of osmotic agents on drying behaviour and product quality. *J. Food Sci.* 50:1217.

Lewicki, P. P., Lenart, A., and Pakula, W. 1984. Influence of artificial semipermeable membranes on the process of osmotic dehydration of apples. Ann. Warsaw Agricult. Univ.-SGGW-AR. *Food Technol. Nutr.* 16:17–24.

Martinez-Monzo, J., Martinez-Navarrete, N, Chiralt, A., and Fito, P. 1998. Mechanical and structural changes in Apple (var. Granny Smith) due to vacuum impregnation with cryoprotectants. *J. Food Sci.* 63(3):499–503.

Mavroudis, N. E., Gekas, V., and Sjoholm, I. 1998a. Osmotic dehydration of apples. Shrinkage phenomena and the significance of initial structure on mass transfer rates. *J. Food Eng.* 38:101–123.

Mavroudis, N. E., Gekas, V., and Sjoholm, I. 1998b. Osmotic dehydration of apples: Effects of agitation and raw material characteristics. *J. Food Eng.* 35:191–209.

Oliveira, F. A. R. and Silva, L. M. 1992. Freezing influences diffusion of reducing sugars in carrot cortex. *J. Food Sci.* 57(4):932.

Ponting, J. D. 1973. Osmotic dehydration of fruits—Recent modifications and applications. *Process Biochem.* 8:18.

Ponting, J. D., Watters, G. G., Forrey, R. R., Jackson, R., and Stanley, W. L. 1966. Osmotic dehydration of fruits. *Food Technol.* 29(10):125.

Rahman, M. S., Perera, C. O., Chen, X. D., Driscoll, R. H., and Potluri, P. L. 1996. Density, shrinkage and porosity of calamari mantle meat during air-drying in a cabinet dryer as a function of water content. *J. Food Eng.* 30:135–145.

Rastogi, N. K. and Niranjan, K. 1998. Enhanced mass transfer during osmotic dehydration of high pressure treated pineapple. *J. Food Sci.* 63(3):508–511.

Simal, S., Benedito, J., Sanchez, E., and C. Rossello. 1998. Use of ultrasound to increase mass transport rates during osmotic dehydration. *J. Food Eng.* 36:323–336.

Torreggiani, D. 1993. Osmotic dehydration in fruit and vegetable processing. *Food Res. Int.* 26:59–68.

Vincent, J. F. V. 1989. Relationship between density and stiffness of apple flesh. *J. Sci. Food Agr.* 47:443–462.

Wienjes, A. G. 1968. The influence of sugar concentration on the vapor pressure of food odor volatiles in aqueous solutions. *J. Food Sci.* 33:1–2.

CHAPTER 5

Influence of Edible Coatings on Osmotic Treatment of Apples

R. DABROWSKA
A. LENART

INTRODUCTION

THE application of coatings as a barrier protects food products from undesirable microbiological, chemical, and physical degradations. Coatings should be biodegradable, able to keep a modified atmosphere around the product, and should act as water-reduction promoters (Wong et al., 1994).

The major disadvantage of osmotic dehydration, limiting its application to foods, is the penetration of osmotic solute inside the material. Coating the food to be dehydrated with an artificial barrier on the surface may efficiently hinder the penetration of solute inside the food, not affecting much the rate of water removal (Ishikawa and Nara, 1993). The properties of coatings depend on their composition and on the method used for the fabrication of coatings. Each coating should be individually examined, its composition determined, and the conditions and restrictions for which it has the required functions should be given (Wong et al., 1994). Aqueous solutions of potato and corn starches, gelatin, amylopectin, pectin, maltodextrin, wheat gluten, sodium alginate, methylcellulose, carboxymethylcellulose, cellulose ethylene, acetylomonoglycerol, chitosan gel, and alcoholic solution of beeswax are used for coating fruit and vegetables (Lewicki et al., 1984; Camirand et al., 1992; Wong et al., 1994).

The aim of this work was to determine the influence of edible coatings on the mass transfer during osmotic treatment of apples, depending on the concentration of carbohydrate solution and drying time of coatings.

43

EXPERIMENTAL

The apples var. Idared were used after being stored for 2–3 months at 5°C and 80–90% relative humidity. Apples were washed, peeled, and diced to 1-cm cubes.

The apple cubes were dipped in the carbohydrate solution at 25°C for 3 min, dried in a drying chamber at 70°C for 10 or 40 min, and dehydrated osmotically in the 61.5% sugar solution (water activity 0.90) at 30°C for 10 or 180 min.

The dry matter content was determined according to the Polish Standard (PS, 1990). The results were analyzed on the basis of water loss (g/g initial d.m.) and solids gain (g/g initial d.m.).

The following carbohydrates were used:

(1) Capsul-E. Producer: National Starch & Chemical Corporation, Kalisz, Poland. Composition: corn starch. Function: low-viscosity, oxidative-resistance improver. Concentration: 3%.
(2) Hi-Flo. Producer: National Starch & Chemical Corporation, Kalisz, Poland. Composition: corn starch. Function: stabilizer, shelf-life extender. Concentration: 3%.
(3) Purity Gum. Producer: National Starch & Chemical Corporation, Kalisz, Poland. Composition: corn starch. Function: emulsion, flavor, and opacifer stabilizer. Concentration: 3%.
(4) Potato starch. Producer: PZP Pleszew, Poland. Composition: amylose and amylopectin. Function: gelatinizing agent. Concentration: 3%.
(5) Maltodextrin. Producer: Szczecin Potato Enterprise, Lobza, Poland. Composition: maltodextrin (DE 22.19). Function: stabilizer, anti-agglomerant. Concentration: 20% and 50%.
(6) High methylated pectin (HMP). Producer: ZPOW Pectowin, Jaslo, Poland. Composition: degree of esthrification 61.9%. Function: gelatinizing agent. Concentration: 3%.
(7) Low methylated pectin (LMP). Producer: Division of Hercules Incorporated, Lille Skensved, Denmark. Composition: degree of esthrification 34%. Function: gelatinizing agent. Concentration: 2%.

RESULTS AND DISCUSSION

The apples coated with 3% solution of Purity Gum or 2% solution of LMP and dried for 10 min had higher dehydration degree after osmotic dehydration for 10 min than other coated and uncoated apples (Figure 5.1).

Also apples coated with 3% Purity Gum, 3% HMP, 2% LMP solution, and 40-min dried had lower water content after osmotic dehydration for 10 min than other coated and uncoated 10-min dried apples (Figure 5.2).

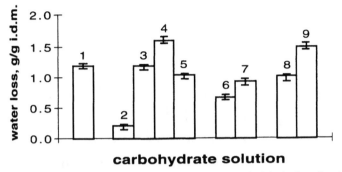

Figure 5.1 Influence of edible coating on water loss during osmotic dehydration of apples. Drying time, 10 min. Osmotic dehydration time, 10 min. 1, uncoated apple. Coated apple: 2, 3% Capsul-E; 3, 3% Hi-Flo; 4, 3% Purity Gum; 5, 3% Potato starch; 6, 20% Maltodextrin; 7, 50% Maltodextrin; 8, 3% HMP; 9, 2% LMP.

Generally, the apples uncoated and coated with the modified starch solutions (Capsul-E, Hi-Flo, and Purity Gum), dried, and 180 min osmotic dehydrated had similar water loss. Only apples coated with 2% LMP solution and dried for 10 or 40 min had the highest dehydration degree after osmotic dehydration for 180 min. Other polysaccharide solutions caused smaller water loss than uncoated 180 min osmotic dehydrated apples (Figures 5.3 and 5.4).

The 10-min dried and 10-min osmotic dehydrated apples coated with 3% Capsul-E, 20% and 50% maltodextrin had smaller solids gain than uncoated 10-min dried apples (Figure 5.5). Other polysaccharide solutions caused bigger solids gain in coated apples than in uncoated apples.

The apples coated with the carbohydrate solutions, excluding those coated with 3% Purity Gum solution, 40-min dried, and 10-min osmotic dehydrated,

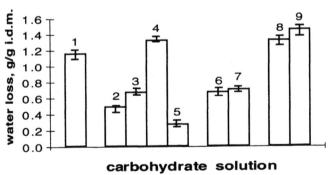

Figure 5.2 Influence of edible coating on water loss during osmotic dehydration of apples. Drying time, 40 min. Osmotic dehydration time, 10 min. 1, uncoated apple. Coated apple: 2, 3% Capsul-E; 3, 3% Hi-Flo; 4, 3% Purity Gum; 5, 3% Potato starch; 6, 20% Maltodextrin; 7, 50% Maltodextrin; 8, 3% HMP; 9, 2% LMP.

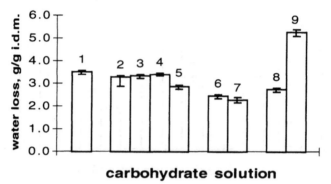

Figure 5.3 Influence of edible coating on water loss during osmotic dehydration of apples. Drying time, 10 min. Osmotic dehydration time, 180 min. 1, uncoated apple. Coated apple: 2, 3% Capsul-E; 3, 3% Hi-Flo; 4, 3% Purity Gum; 5, 3% Potato starch; 6, 20% Maltodextrin; 7, 50% Maltodextrin; 8, 3% HMP; 9, 2% LMP.

were observed to have smaller or similar solids gain than uncoated (Figure 5.6).

The apples uncoated and coated with starch solutions (modified starch and potato starch), dried 10 min and 180 min osmotic dehydrated had similar solids gain (Figure 5.7). Other carbohydrate solutions caused smaller solids gain in coated apples than in uncoated apples.

The 40-min dried and 180 min osmotic dehydrated coated apples had smaller solids gain than uncoated 180 min osmotic dehydrated apples independently on edible coatings (Figure 5.8).

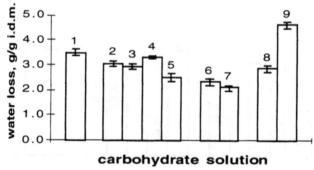

Figure 5.4 Influence of edible coating on water loss during osmotic dehydration of apples. Drying time, 40 min. Osmotic dehydration time, 180 min. 1, uncoated apple. Coated apple: 2, 3% Capsul-E; 3, 3% Hi-Flo; 4, 3% Purity Gum; 5, 3% Potato starch; 6, 20% Maltodextrin; 7, 50% Maltodextrin; 8, 3% HMP; 9, 2% LMP.

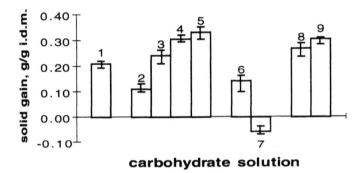

Figure 5.5 Influence of edible coating on solids gain during osmotic dehydration of apples. Drying time, 10 min. Osmotic dehydration time, 10 min. 1, uncoated apple. Coated apple: 2, 3% Capsul-E; 3, 3% Hi-Flo; 4, 3% Purity Gum; 5, 3% Potato starch; 6, 20% Maltodextrin; 7, 50% Maltodextrin; 8, 3% HMP; 9, 2% LMP.

CONCLUSIONS

Water loss for coated apples were lower than for uncoated ones. Only apples coated with 2% LMP solution had the highest dehydration degree and was not dependent on the drying time (10 or 40 min) and osmotic dehydration time (10 or 180 min). LMP coating was chosen as the best one, considering moisture losses.

The penetration of osmotic solute into coated apples was generally reduced, or it was on the comparable level with uncoated apples, depending on the dry-

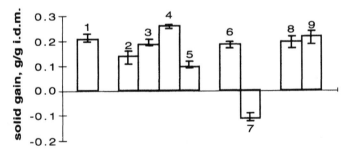

Figure 5.6 Influence of edible coating on solids gain during osmotic dehydration of apples. Drying time, 40 min. Osmotic dehydration time, 10 min. 1, uncoated apple. Coated apple: 2, 3% Capsul-E; 3, 3% Hi-Flo; 4, 3% Purity Gum; 5, 3% Potato starch; 6, 20% Maltodextrin; 7, 50% Maltodextrin; 8, 3% HMP; 9, 2% LMP.

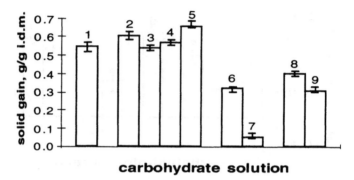

Figure 5.7 Influence of edible coating on solids gain during osmotic dehydration of apples. Drying time, 10 min. Osmotic dehydration time, 180 min. 1, uncoated apple. Coated apple: 2, 3% Capsul-E; 3, 3% Hi-Flo; 4, 3% Purity Gum; 5, 3% Potato starch; 6, 20% Maltodextrin; 7, 50% Maltodextrin; 8, 3% HMP; 9, 2% LMP.

ing time and water loss. Solids gain in 40-min dried apples decreased more after osmotic dehydration than in 10-min dried apples. Solids gain in coated apples increased with the time of osmotic dehydration. Maltodextrin solution was chosen as the best one, considering solids gain.

ACKNOWLEDGEMENT

This work was supported by the State Committee for Scientific Research, grant no. 5P06F 030 16.

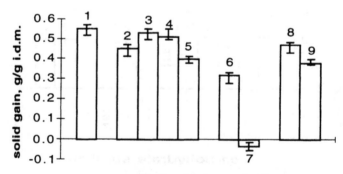

Figure 5.8 Influence of edible coating on solids gain during osmotic dehydration of apples. Drying time, 40 min. Osmotic dehydration time, 180 min. 1, uncoated apple. Coated apple: 2, 3% Capsul-E; 3, 3% Hi-Flo; 4, 3% Purity Gum; 5, 3% Potato starch; 6, 20% Maltodextrin; 7, 50% Maltodextrin; 8, 3% HMP; 9, 2% LMP.

REFERENCES

Camirand, W., Krochta, J. M., Pavlath, A. E., Wong, D., and Cole, M. E. 1992. Properties of some edible carbohydrate polymer coatings for potential use in osmotic dehydration. *Carbohydrate Polymers* 17:39–49.

Ishikawa, M. and Nara, H. 1993. Osmotic dehydration of food by semipermeable membrane coating. In: *Advances in Food Engineering* Singh, R. Paul and Wirakartakusuman, M. A. (Eds). London: CRC Press, p. 73–77.

Lewicki, P. P., Lenart, A., and Pakua, W. 1984. Influence of artificial semi-permeable membranes on the process of osmotic dehydration of apples. Ann. Warsaw Agricult. Univ.-SGGW-AR. *Food Technol. Nutr.* 16:17–24.

Wong, W. S., Tillin, S., Hudson, J. S., and Pavlath, E. 1994. Gas exchange in cut apples with bilayer coatings. *J Agr. Food Chem.* 10(42):2278–2285.

PART II

VACUUM IMPREGNATION AND OSMOTIC PROCESSES IN FRUIT AND VEGETABLES

CHAPTER 6

Vacuum Impregnation Viability of Some Fruits and Vegetables

A. ANDRÉS
D. SALVATORI
A. ALBORS
A. CHIRALT
P. FITO

INTRODUCTION

VACUUM impregnation (VI) of a porous food with an external solution is an interesting treatment to promote fast compositional changes by the action of the hydrodynamic mechanism (HDM). Salting process, osmotic dehydration, and other solid-liquid operations can be improved by applying vacuum pulses to the system, with a significant decrease in the processing time, depending on the product effective porosity and mechanical properties. There is a fast mass transfer mechanism, called HDM, that occurs when porous structures are immersed in a liquid phase. This involves the inflow of the external liquid throughout the capillary pores, controlled by the expansion/compression of the internal gas. This mechanism is responsible for the VI processes of porous products when low pressures are imposed in a solid-liquid system (vacuum step) followed by the restoration of atmospheric pressure. During the vacuum step, the internal gas in the product pores of the product is expanded and partially flows out. All this is coupled with the capillary penetration as a function of the interfacial tension of the liquid and the diameter of pores. In the atmospheric step, the residual gas is compressed, and the external liquid flows into the pores as a function of the compression ratio (Fito and Pastor, 1994). Nevertheless, pressure changes can also promote deformations of the product because of the viscoelastic properties of its solid matrix. Coupling of HDM with the deformation-relaxation phenomena (DRP) of the product's solid matrix has been described and modeled (Andrés, 1996; Fito et al, 1996), showing that volume changes at the end of the vacuum and the atmospheric steps, as well as the effective porosity affect the volume fraction of the product impregnated by the external liquid.

VI can improve the mass transfer rate in many processes where solid-liquid operations are involved: salting, osmotic dehydration, acidification, addition of preservatives, etc. (Fito and Chiralt, 1995; Shi et al., 1996; Chiralt and Fito, 1997; Fito and Chiralt, 1997; Chiralt et al., 1999). The VI technique allows us to introduce the desired food ingredient directly into the product throughout its pores, in a controlled way, according to the HDM model. However, knowledge of the product porosity and the feasibility of liquid penetration during the VI operation are required to apply it adequately. Fruits and vegetables have a great part of their internal volume occupied by gas (Baumann and Henze, 1983; Calbo and Sommer, 1987). In fruit processing, such as osmotic dehydration or minimal processing, VI allows fast compositional changes to be made by introducing appropriate solutions (with water activity and pH depressors, preservatives, etc.) into their porous structure.

The aim of this work is to study the coupling of HDM and DRP in some fruits to determine the feasibility of applying VI to modify fruit composition and structure for a number of industrial uses (minimal processing, freezing pretreatments, etc.). In this paper, the behavior of some fruits of industrial interest throughout VI process has been analyzed. The effective porosity (ϵ_e), the volume fraction of the impregnated liquid (X), and the relative deformations (γ) of fruit samples, because of the pressure changes, were evaluated as a function of the pressure ratio and the pathway of the applied treatment.

CHARACTERISTIC PARAMETERS OF VACUUM IMPREGNATION

VI properties can be determined by using previously described equipment (Fito et al., 1996; Salvatori et al., 1997, 1998), which allows us to measure the sample mass and volume changes throughout the VI process. From these kind of data, it is possible to calculate the VI characteristic parameters: impregnated sample volume fraction (X), sample volume deformation (γ), and the effective porosity (ϵ_e) of the sample to the HDM action (Fito et al., 1996). The effective porosity is related with the other characteristic parameters by the model equation proposed by Fito et al. (1996):

$$\epsilon_e = \frac{(X - \gamma)r + \gamma_1}{r - 1} \tag{1}$$

where:

γ_1 = relative sample volume deformation at the end of the vacuum period (m^3 of sample deformation/m^3 of sample at $t = 0$)

γ = final relative volume sample deformation (m^3 of sample deformation/m^3 of sample at $t = 0$)

X = volumetric fraction of sample occupied by liquid as a result of HDM at the end of the process (m^3 of liquid/m^3 of sample at $t = 0$)

ϵ_e = effective porosity (m^3 of gas inside the pores/m^3 of sample)

r = compression ratio ($\sim$ atmospheric pressure/vacuum pressure)

The volume fraction of sample impregnated by liquid after the vacuum step X_1 (m^3 of liquid/m^3 of sample at $t = 0$) is another characteristic parameter that can be calculated from the experimental data (Salvatori et al., 1998).

These parameters are all related because both mechanisms (impregnation and deformation) are coupled. The viscoelastic nature of the fruit tissue and the pressure gradients imposed on the system during the process are the reasons for the coupling of impregnation and deformation mechanisms.

The impregnation and deformation levels of the fruits can be measured at the end of the vacuum step (X_1 and γ_1) and at the end of the VI process (X and γ). Table 6.1 shows the values of these characteristic parameters for different kinds of fruits (Salvatori et al., 1998; Sousa et al., 1998; Fito and Chiralt, 2000). The fast kinetic of vacuum impregnation of fruits is an important characteristic of this mass transfer mechanism. A few minutes (5–15 min, depending on the size and porosity of the fruit) are enough to degasify the structure.

The values of γ_1 observed for each product show that the solid matrix deformation (volume swelling) occurred because of the internal gas expansion during vacuum period. Negative values of impregnation in the vacuum step (X_1) must be due to the losses of native liquid carried away by the gas expansion and flow out from the pores. Great negative X_1 values have been obtained in apple (*G. Smith* and *R. Chief*), pineapple, melon, banana, eggplant, and orange peel, whereas the fruit that expanded most during the vacuum step were mango and kiwifruit.

The values of γ represent the net volume change at the end of the VI process, resulting from an initial swelling throughout the vacuum step and the later compression in the second VI process, both of them coupled with the impregnation phenomena. It is remarkable that orange peel and mango achieve a notable final swelling, which implies an increase in the pore volume useful for impregnation. Swelling (to a minor extent) also occurs in peach (*Miraflores*), pineapple, pear, and zucchini. Eggplant showed the highest volume reduction, and apple (*Golden*), peach (*Catherine*), and strawberry also are compacted. In the other fruits, no significant volume change or volume compression was observed.

The X values represent the net impregnation level at the end of the VI process. However, if a negative value of X_1 is achieved, the total external liquid penetrated into the fruit during VI will be $X + X_1$, and this sum represents the actual effectiveness of VI to promote compositional changes using an external solution of controlled composition. So, in Table 6.1, the actual im-

TABLE 6.1. Mean Values of VI Parameters Obtained for Different Fruits.

Fruit Variety	Sample	X_i	γ_i	X	γ	ϵ_e	ϵ_e/ϵ_r	X_{LN}^a
Apple G. Smith	2 × 2 cm cylinders	-4.2	1.7	19.0	-0.6	21.0	0.88	8.5
Apple R. Chief	2 × 2 cm cylinders	-5.0	2.1	17.9	-2.4	20.3	0.94	10.6
Apple Golden	2 × 2 cm cylinders	-2.7	2.8	11.2	-6.0	17.4	0.69	7.5
Mango Tommy Atkins	1-cm slices	0.9	5.4	14.2	8.9	5.9	0.60	0.73
Strawberry Chandler	Whole fruit	-2.1	2.9	1.9	-4.0	6.4	1.02	3.4
Kiwi-fruit Miraflores	Fruit quarters	-0.2	6.8	1.09	0.8	0.66	0.29	0 34
Peach Hayward	2.5-cm cube sides	-2.29	2.0	6.5	2.1	4.7	1.81	—
Peach Catherine	2.5-cm cube sides	-1.4	0.5	4.4	-4.2	9.1	1.14	2.2
Apricot Bulida	Fruit halves	-0.2	1.5	2.1	0.11	2.2	1.11	0.84
Pineapple Espanola Roja	1-cm slices	-6.5	1.8	5.7	2.3	3.7	1.85	8.0

(continued)

TABLE 6.1. (continued).

Fruit Variety	Sample	X_i	γ_i	X	γ	ϵ_e	ϵ_e/ϵ_r	X_{LN}[a]
Pear Passa Crasana	2.5-cm cube sides	−1.3	2.8	5.3	2.2	3.4	0.68	1.6
Prune President	Fruit halves	−1.0	0.6	1.0	−0.8	2.0	1.13	1.4
Melon Inodorus	2 × 2 cm cylinders	−4.0	2.0	5.0	−0.4	6.0	1.40	5.1
Banana Giant Cavendish	Slices 0.5–2 cm height	−6.1	3.6	10.6	1.3	10.1	—	8.9
Eggplant	1 × 1 cm cubes	−9.6	2.4	15	−37	53.8	0.93	24.4
Zucchini	2.5-cm cube	−2.47	3.2	5.85	3.27	2.6	0.15	3.2
Orange peel Navel late	2.5 × 6 cm	−6	5	40	14	20	0.72	11.5

[a]Volume fraction (percentage) of native liquid lost throughout the vacuum step.

pregnation $(X + X_1)$ is shown as X values. Orange peel is the most impregnated product followed by the apples and mango. These results reflect the suitability of these products to mass transfer operation using the vacuum technology. No significant net impregnation was reached in strawberry, but taking the X_1 value into account 2% (v/v) of liquid may be replaced. In pineapple and melon, the net impregnation level was similar to X_1; this means that the native liquid was exchanged for the external solution.

The effective porosity values (ϵ_e) were estimated by Equation (1) by substituting the actual impregnation level (Table 6.1). So, these values include not only the pore volume occupied by gas but also by replaced native liquid.

The ratio effective porosity/actual porosity (ϵ_e/ϵ_r) is also a noteworthy parameter. This ratio (ϵ_e/ϵ_r) represents the porous fraction of the fruit that is available to the HDM action. This means that for most of the studied fruits, between 69–95% of the porous fraction can be filled by liquid after a vacuum impregnation treatment at 50 mbar. But there are some other products with $\epsilon_e/\epsilon_r > 1$, just those with a great part of their intercellular spaces (pores) occupied by native liquid as indicated by the values of X_1.

In the vacuum step, two opposite fluxes occur in fruit pores, related with the native liquid outflow $(X_{LN} < 0)$ and the capillary inflow of external solution $(X_c > 0)$. So, the X_{LN} values can be estimated by Equation (2) at the different values of p_1 applied during VI process, if X_c is known or can be calculated as shown in Equation (3).

$$X_{LN} = X_c - X_1 \tag{2}$$

$$X_c = \epsilon_e \, (p_c/(p_1 + p_c)) \tag{3}$$

For *G. Smith* apple, p_c can be estimated to be 19 mbar, taking into account the average pore diameter (160 μm) reported by Fito and Pastor (1994). Table 6.1 shows the mean X_{LN} value for *G. Smith* apple and for other studied products. This parameter was calculated by considering the same capillary pressure in all cases as estimated values.

INFLUENCE OF VACUUM PRESSURE

Table 6.2 shows an example of data obtained at different p_1 values for *G. Smith* apple. As the theoretical model predicts, coupling of the deformation and penetration phenomena occurs to a different extent in both VI steps (vacuum and atmospheric ones), depending on pressure ratio (impregnation) or difference (deformation) and sample mechanical and structural properties. This fact makes the observation of a clear effect of vacuum pressure (p_1) on each parameter (γ or X) difficult, so the statistical analysis (ANOVA) did not

TABLE 6.2. Influence of Vacuum Level on the VI
Parameters Obtained for *Granny Smith* Apple.

Apple (var. *Granny Smith*)					
$p1$ (mbar)	X_i	γ_i	X	γ	ϵ_e
50	−4.6	1.1	19.5	−0.8	19
100	−4.7	1.7	18.2	−1.3	17
200	−6.3	1.3	22.6	0.3	19
400	−8.2	1.0	22.6	0.7	17
Mean value	a	1.3	20.7	−0.3	18.2

[a]Significant differences among samples due to working pressure ($\alpha < 0.05$).

reflect a significant influence of vacuum pressure on the VI parameters in most of the cases. Nevertheless, in all the experiments similar values of ϵ_e were estimated for each fruit, which validates the coupled model (Fito et al., 1996).

Only the X_1 values appeared clearly dependent on p_1; the higher the p_1 value the more negative the X_1. This was in agreement with the level of capillary penetration expected from Equation (3) (Fito, 1994). It seems to indicate that by using p_1 lower than 400 mbar it is possible to remove practically all the native liquid from the pore structure and the capillary penetration (pressure dependent) is the flux that marks differences between X_1 values.

CONCLUSIONS

The volume fraction of fruit impregnated by an external solution is the most relevant parameter when VI is used to promote compositional changes (e.g., to improve stability or quality of a minimally processed fruit). In rigid porous solids, this fraction depends only on porosity and pressure ratio. Nevertheless, in viscoelastic fruits, coupling of deformation-impregnation phenomena and the substitution of native liquid for the external solution greatly affect the final amount of the impregnated external liquid. So, all these effects must be taken into account to evaluate the feasibility of VI in different fruits and vegetables.

REFERENCES

Andrés, A. 1996. *Impregnación a vacío en alimentos porosos. Aplicación al salado de queso.* Ph.D. thesis, Polytechnic University of Valencia.

Baumann, H. and Henze, J. 1983. Intercellular space volume of fruit. *Acta Hort* 138:107–111.

Calbo, A. G. and Sommer, N. F. 1987. Intercellular volume and resistance to air flow of fruits and vegetables. *J. Am. Soc. Hort. Sci.* 112(1):131–134.

Chiralt, A. and Fito, P. 1997. Salting of manchego-type cheese by vacuum impregnation. In Fito, P., Ortega, E., and Barvosa, G. (Eds.). *Food Engineering 2000.* New York: Chapman & Hall, pp. 215–230.

Chiralt, A., Fito, P., Andres, A., Barat, J. M., Martinez-Monzó, J., and Martinez-Navarrete, N. 1999. Vacuum impregnation: a tool in minimally processing of foods. In Oliveira, F. A. R. and Oliveira, J. C. (Eds.). *Processing of Foods: Quality Optimization and Process Assessment.* Boca Raton: CRC Press, pp. 314–356.

Fito, P. 1994. Modelling of vacuum osmotic dehydration of foods. *J. Food Eng.* 22:313–318.

Fito, P. and Chiralt, A. 1995. An update on vacuum osmotic dehydration. In Barbosa-Cánovas, G. V. and Welti-Chanes, J., (Eds.). *Food Preservation by Moisture Control: Fundamentals and Applications.* Lancaster, PA: Technomic Publishing Co., Inc., pp. 351–372.

Fito, P. and Chiralt, A. 1997. An approach to the modelling of solid food-liquid operations: Application to osmotic dehydration. In Fito, P., Ortega, E., and Barvosa, G. (Eds.). *Food Engineering 2000.* New York: Chapman & Hall, pp. 231–252.

Fito, P. and Chiralt, A. 2000. Vacuum impregnation of plant tissues. In Alzamora, S. M., Tapia, M. S., and Lopez-Malo, A. (Eds.). *Minimally Processed Fruits and Vegetables.* Gaithersburg, MD: Aspen Publishers, Inc., pp. 189–204.

Fito, P. and Pastor, R. 1994. On some non-diffusional mechanism occurring during vacuum osmotic dehydration. *J. Food Eng.* 21:513–519.

Fito, P., Andrés, A., Chiralt, A., and Pardo, P. 1996. Coupling of hydrodynamic mechanism and deformation-relaxation phenomena during vacuum treatments in solid porous food-liquid systems. *J. Food Eng.* 27:229–240.

Salvatori, D., Andrés, A., Chiralt, A., and Fito, P. 1998. The response of some properties of fruits to vacuum impregnation. *J. Food Process Eng.* 21:59–73.

Salvatori, D., Da Silva, J., Andres, A., Chiralt, A., and Fito, P. 1997. Caracterización del acoplamiento de los fenómenos deformación-penetración en frutas, durante su impregnación a vacío. In Ortega, E., Parada y, E., Fito, P. (Eds.). *Equipos y procesos para la industria alimentaria.* Valencia: Servicio de Publicaciones, Universidad Politécnica de Valencia, pp. 357–363.

Shi, X. Q., Chiralt, A., Fito, P., Serra, J., Escoin, C., and Gasque, L. 1996. Application of osmotic dehydration technology on jam processing. *Drying Technol.* 14(3&4):841–847.

Sousa, R., Salvatori, D., Andrés, A., and Fito, P. 1998. Analysis of vacuum impregnation of banana (*Musa acuminata cv. Giant Cavendish*). *Food Sci. Technol. Int.* 4:127–131.

CHAPTER 7

Combined Vacuum Impregnation-Osmotic Dehydration in Fruit Cryoprotection

J. MARTÍNEZ-MONZÓ
N. MARTÍNEZ-NAVARRETE
A. CHIRALT
P. FITO

INTRODUCTION

PRESERVATION of food by freezing is a good method of ensuring the long-term retention of original characteristics, in almost unchanged state, especially of perishable materials. This quality retention is achieved by the combined effect of the low temperature, slowing down both biochemical and microbial activities, and the decrease in the water chemical potential (ice formation) (Instituto Internacional del Frío, 1990). Freezing of fruit results in various favorable effects with respect to the shelf life and availability throughout the year; nevertheless, various undesirable changes occur because of this process. Freezing destroys cell integrity and compartmentation, thereby increasing the opportunity of undesirable physical, chemical, and biochemical reactions (browning, texture changes, loss of flavor, etc.) (Eskin, 1989; Roos, 1993). The consequence is, that during frozen storage, a gradual cumulative and irreversible loss of quality occurs in time. Prefreezing treatments, selection of the optimum freezing rate, adequate packaging, correct and uniform storage temperature, and rate of subsequent thawing are crucial if the full benefits of food freezing have to be realized and the deteriorative reactions minimized.

Cryostabilization technology represents a conceptual approach to a practical industrial technology for the stabilization during processing and storage of frozen foods (Slade and Levine, 1991). The key to cryoprotection lies in controlling the physical state of the freeze-concentrated amorphous matrix surrounding the ice crystals in a frozen system, where deteriorative reactions mainly occur. There are two possibilities for achieving an adequate food cryoprotection (Slade and Levine, 1995). One is the reduction in the water content of the product below W_g' (content of unfrozen water in the frozen product), allowing its complete vitrification. The technique is termed *dehydrofreezing* and the concentration step is generally realized by air drying (Torreggiani et al., 1987), osmotic dehydration (Forni et al., 1987; Pinnavaia et al., 1988; Giangiacomo et al., 1994); or a combination of both (Robbers et al., 1997). Another is the formulation of food with appropriate ingredients to elevate T_g' relative to freezer temperature, thereby enhancing the product stability (Levine and Slade, 1989).

The specific role of some solutes in protecting cell membranes during cell water loss in drying or cryoconcentration during freezing has been reported (Torregiani, 1995; Crowe et al., 1998). Nevertheless, the possibility of introducing solutes into structured food such as fruit is not easily feasible. Vacuum impregnation technique (Fito, 1994; Fito and Chiralt, 1995) can offer interesting prospects in developing pretreatments to modify (in a short time) the initial composition of porous fruits, introducing cryoprotectant solutes and making them more suitable for resisting damages caused by the frozen-thawing processes. If the impregnation solution is hypertonic, cryostabilization is obtained by using the combination of vacuum impregnation and osmotic dehydration. In addition, some benefits from the reduction of the amount of oxygen inside food pores, such as a greater stability against some deteriorative reactions (browning and oxidations), can be obtained.

In this chapter, the effectiveness of vacuum impregnation (VI) in modifying composition and structure of porous fruits with different kinds of solutes (high and low molecular weight), combined with water removal, is discussed, as well as their influence on some physical and structural properties. Likewise, the possibility to enhance product stability during freezing-thawing processes is analyzed on the basis of the changes promoted in the water behavior in non-equilibrium state at subzero temperatures. Special emphasis is placed on apple, because of its highly porous structure and, therefore, the great potential of VI (Martínez-Monzó et al., 1998a; Salvatori et al., 1998) to promote changes in the product.

FAST CHANGES OF FRUIT COMPOSITION BY VI

In the past, vacuum pressure has been applied in several processes with differing aims: the minimal processing of fruit to incorporate different additives (del Rio and Miller, 1979; Santerre et al., 1988), whey and air removal in curd cheddar cheese (Reinbold, 1993), and osmotic dehydration processes (Zozulevich and D'yachenco, 1969; Hawkes and Flink, 1978). However, no clear reasons explaining the role of vacuum in the different processes were reported until 1994, when Fito explained the action of vacuum pressure in porous products immersed in a liquid phase by trying to model the faster kinetics of vacuum osmotic dehydration of apples (Fito, 1994). In a VI process, a porous product is immersed in an adequate liquid phase and is submitted to a two-step pressure change. First, a vacuum pressure promotes the gas flow throughout the porous product until mechanical equilibrium is achieved. In this moment, capillary penetration will be higher than at atmospheric pressure. When atmospheric pressure is restored in a second step, residual gas compression leads to the external solution inflow while pressure gradients persist. This phenomenon was called the hydrodynamic mechanism (HDM). From this point of view, the observed acceleration of the osmotic process was justified by the effect of pressure gradients that promote an overlapped hydrodynamic flow into the pores, coupled with diffusional-osmotic mechanisms. From the HDM model, it is possible to predict the amount of liquid that can be introduced into a porous food (Fito and Chiralt, 1995), with differing aims: modifying the composition introducing additives, salting, etc., and it is also possible to evaluate the effectiveness of a vacuum treatment in expelling internal gas or liquid. The volume fraction of the initial sample (X) impregnated by the external liquid when mechanical equilibrium is achieved in the sample has been modeled in a simplified way for stiff products, as a function of the compression ratio r (r = atmospheric pressure (p_2)/vacuum pressure (p_1)), and sample effective porosity (ϵ_e), as described in Equation (1) (Fito, 1994). Characteristic times of VI of porous foods are very short, although some factors such as the impregnating solution viscosity may affect kinetics of the impregnation process and its coupling with deformation of viscoelastic samples because of pressure gradients (Chiralt et al., 1999).

$$X = \epsilon_e \left[1 - p_1/p_2 \right] \tag{1}$$

From this point of view, VI may be a useful tool to promote fast compositional changes, including water content reduction, if low water activity solutions are used for VI. In this sense, in osmotic dehydration processes of porous fruits, VI pretreatment can be performed with the osmotic solution at the beginning of the process. This is called pulsed vacuum osmotic dehydration (PVOD) and has been observed to be very effective in promoting mass transfer kinetics (Chiralt et al., 1999; Shi and Fito, 1993; Shi et al., 1995; Fito and Chiralt, 1997). VI implies an external solute and water transport inside the tissue while modifying the effective diffusion coefficient in the product liquid phase (PLP). The gas-liquid exchange implies a fast change in the sample overall composition that modifies the process-driven force at the very beginning of the process while pores still have liquid. Effective diffusion promotion by VI in the fruit liquid phase is especially important when the impregnated solution has a low viscosity (Martínez-Monzó et al., 1998b). This fits in with the coupling of diffusional and osmotic (water selective diffusion though cell membrane) mechanisms to a different extent in each case. If a sugared solution is used for VI, sugar gain occurs through diffusion in the pore, whereas this mechanism plays a minor role in water loss that occurs mainly cell to cell by osmotic mechanism. When intercellular spaces are occupied by gas (non-impregnated sample) or by a very viscous solution, solute diffusion through intercellular spaces seems to be inhibited and the sample reaches the chemical equilibrium with the osmotic solution with greater osmotic water loss. This aspect also implies greater product liquid phase and volume losses of the samples.

Table 7.1 reflects compositional changes induced by VI and/or osmotic dehydration in cylindrical apple samples (2 cm height and diameter). HM pectin is introduced into the pores of apple samples, as a possible cryostabilizer, as well as low molecular weight sugars (glucose and fructose) from rectified-concentrated grape must (GM) as cryoprotectants. Sample water reduction by osmotic treatments with GM was also performed. Three sample groups, each one with a water activity level (0.985, 0.970, and 0.955) were obtained, each one containing samples with different components in the intercellular spaces (air, high molecular weight solutes, or low molecular weight solutes). The group with higher a_w was the untreated samples (F), and those impregnated with an isotonic pectin solution (IP). The second and third a_w levels were obtained by applying three different kinds of treatments: (a) VI of the samples with 35 and 61°Brix GM (sample IM1 and IM2, respectively), (b) VI of sample with isotonic pectin solution, followed by osmotic dehydration in 61°Brix GM till the respective a_w level (samples IP-OD1 and IP-OD2), and (c) osmotic dehydration in 61°Brix at atmospheric pressure with any previous impregnation pretreatment (OD1 and OD2 samples). VI and osmotic treatments were performed at 20°C. In the former, samples were immersed in the corresponding solution for 30 min (15 min at 50 mbar followed by 15 min at atmospheric pressure). IP solution contained 2% w/w of polymer in adequately diluted GM.

Changes in physicochemical properties, color, and mechanical properties induced by each kind of treatment, as affected by the pore constituents of the

TABLE 7.1. Sample Composition (Water and Solute Mass Fraction and Water Activity.

Sample	x_w	x_s	a_w
F	0.84 ± 0.01	0.14 ± 0.01	0.986 ± 0.003
OD1	0.73 ± 0.02	0.24 ± 0.02	0.973 ± 0.004
OD2	0.65 ± 0.02	0.31 ± 0.03	0.955 ± 0.006
IP	0.84 ± 0.01	0.14 ± 0.01	0.982 ± 0.003
IP-OD1	0.75 ± 0.01	0.22 ± 0.01	0.970 ± 0.004
IP-OD2	0.68 ± 0.01	0.29 ± 0.01	0.958 ± 0.005
IM1	0.79 ± 0.01	0.19 ± 0.00	0.972 ± 0.004
IM2	0.68 ± 0.01	0.29 ± 0.02	0.957 ± 0.006

sample and their a_w reduction, are analyzed in the following paragraphs. Likewise, their influence on changes promoted by freezing-thawing in the samples is also discussed.

CHANGES IN MECHANICAL, STRUCTURAL, AND COLOR PROPERTIES

Structured foods such as fruits and vegetables have good aptitudes to VI because of their porosity, to enhance diffusional mechanisms in dehydration or to develop new products (modifying their original composition) without disrupting their cellular structure. Nevertheless, it has been observed that VI has a different effect on some physical properties (Chiralt et al., 1999). Structural and mechanical properties of the product can be affected because of pressure changes and the replacement of native liquid or air in the pores by an external solution. In treatments with hypertonic solution, loss of cell turgor will also greatly affect mechanical behavior (Martínez-Monzó et al., 1998b; Pitt, 1992). On the other hand, because VI implies a greater homogeneity of sample refraction index due to the air-liquid exchange, optical and color properties will also change (Fito et al., 2000).

For mechanical properties, a greater elastic behavior of samples containing air in the pores could be expected when no turgor loss occurs (Martínez-Monzó et al., 1998b). On the other hand, presence of pectin could contribute to increasing the cohesiveness of cellular structure, thus also increasing sample elasticity. For samples described in Table 7.1, stress relaxation tests (8% constant strain, 200 mm/min deformation rate) were performed to analyze mechanical behavior. The obtained stress relaxation curves are plotted in Figure 7.1 where only one curve appears for samples with similar behavior: samples OD1 and IP-OD1 and samples IM1, IM2, OD2, and IP-OD2. Curves were analyzed in terms of the initial (E_i) (Rao, 1992) and asymptotic (E_a) elastic modulus and A (relative relaxation level) and B (relaxation rate) parameters from Peleg model (Peleg, 1980; Peleg and Pollak, 1982). Their values appear in Table 7.2.

When IP solution was used for VI (cell turgor unaltered), no significant differences between the initial elasticity modulus of fresh and IP apples were found. Nevertheless, the relaxation rate and the total relaxation level slowly increased in VI sample. In a previous article (Martínez-Monzó et al., 1996), this increase has been related to the flow of the impregnated free liquid in the intercellular spaces. On the other hand, sample dehydration implied sharp changes in their mechanical response, the greater the a_w reduction the greater the changes in mechanical parameters. The loss of cell turgor due to osmotic treatments or VI with hypertonic solutions led to a steep decrease in stress-strain relationships: samples became "less elastic-more viscous" as deduced

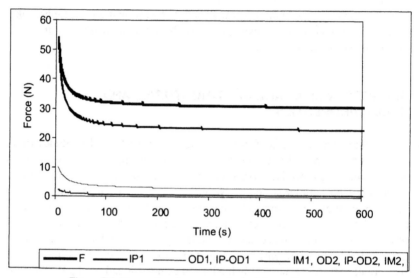

Figure 7.1 Stress relaxation curves obtained for studied samples.

from the increase in the relaxation rate (B) values and the relative relaxation levels (A).

Nevertheless, for a given a_w level, E_a and E_i values are greater for OD and IP-OD samples. This finding agrees with that commented on above related to the gas contribution to the elastic behavior and with possible structural benefits of the impregnated pectin.

Cryo-SEM technique allowed us to see how effective the VI technique was in introducing solutes into the apple pores (Martínez-Monzó et al., 1998a; Chiralt et al., 1999). Intercellular spaces appeared empty in fresh samples and completely flooded in ones impregnated with IP or GM. Moreover, the different aspect of the dentritic forms for the intra- and extracellular liquid phase

TABLE 7.2. Effect of Vacuum Impregnation and/or Osmotic Dehydration on Stress Relaxation Parameters of Samples.

Sample	$E_i \cdot 10^{-5}$ (Pa)	$E_a \cdot 10^{-5}$ (Pa)	A	$B(s^{-1})$
F	20 ± 4	12	0.41 ± 0.04	0.13 ± 0.04
OD1	5.2 ± 0.1	1.4	0.74 ± 0.01	0.08 ± 0.01
OD2	1.6 ± 0.8	0.43	0.73 ± 0.02	0.18 ± 0.02
IP	19 ± 3	8	0.58 ± 0.01	0.15 ± 0.02
IP-OD1	7.2 ± 0.5	1.5	0.80 ± 0.01	0.11 ± 0.01
IP-OD2	1.5 ± 0.6	0.38	0.73 ± 0.01	0.10 ± 0.01
IM1	1.0 ± 0.1	0.18	0.83 ± 0.04	0.23 ± 0.04
IM2	0.9 ± 0.5	0.15	0.83 ± 0.06	0.30 ± 0.10

in these samples was also attributed to the different composition of each phase and its influence on the ice microcrystal formation rate. On the other hand, Cryo-SEM observations of isotonic VI samples did not show cellular alterations (size, shape, and intracellular arrangement) or debonding. These results led to the observed changes in the viscoelastic behavior of samples VI with isotonic solutions that can be principally attributed to the exchange of gas (compressible during the mechanical test) for liquid that will flow out from the pores throughout compression (Martínez-Monzó et al., 1996).

In osmosed samples, VI leads to a structural development of the tissue different from that which occurred in non-impregnated samples because of the substitution of air by an impregnating solution. Differences in the structural features observed in vacuum impregnated and non-impregnated samples throughout water loss have been explained in terms of the different drop in the flow pressure of fluid in the intercellular spaces toward the generated volume in the cell when water flows out. This pressure drop is much greater for liquid phases than for gas, which implies the generation of a different force balance on each side of the double-layer plasmalemma-cell wall in line with water loss (Fito et al., 2000). As a consequence, when there is liquid in the pores, plasmalemma separates from the cell wall, and external liquid passes through the cell wall, flooding the cell cavity, with scarce deformation of the cell wall (Martínez-Monzó et al., 1998a). When there is gas in the intercellular space, plasmalemma shrinks together with cell wall that becomes greatly deformed when osmotic process progresses (Barat et al., 1999). The inflow of liquid from intercellular spaces to the inner part of the cell cavity when plasmalemma and cell wall become separated in VI samples could contribute to the establishment of solution solutes and cell wall interactions that could preserve cell wall structure and so the mechanical properties of tissues. The better preservation of structure and mechanical properties of VI-osmosed samples compared with non VI-osmosed samples has been observed for several fruits (Alzamora and Gerschenson, 1997).

VI treatments imply the total or partial substitution of the air present in the pores by the impregnation solution, thus changing the optical properties of the product. Color changes associated with compositional changes by VI-OD were evaluated by measuring CIE $L^*a^*b^*$ coordinates (Illuminant D65/observer 10°) with a Minolota MC 1000 spectrocolorimeter. For samples described in Table 7.1, comparison of the obtained coordinate values (Table 7.3) allows us to observe very small changes due to sample dehydration without VI. Nevertheless, VI treatments lead to lower values of L^*, associated with the commented transparency gain due to air loss. Moreover, VI samples showed a slightly redder, less yellow hue, especially in samples VI with hypertonic solutions, probably due to the different color of GM used as impregnating solution. Presence of pectin causes opacity recovery and reflectance increase.

When total color change induced by each treatment with respect to the non-treated sample is evaluated throughout ΔE parameter, small changes are as-

TABLE 7.3. Color Coordinates, Hue, Chrome, and Color Difference due to Samples' Treatment.

	L^*	a^*	b^*	h^*	C^*	ΔE[a]
F	70 ± 1	−3.2 ± 0.7	15 ± 2	101.6	15.7	0
OD1	67 ± 1	−3.6 ± 0.8	18 ± 2	101.2	18.3	3.7
OD2	66 ± 2	−3.0 ± 1.2	20 ± 3	98.7	19.9	5.9
IP	36 ± 1	0.0 ± 0.5	7 ± 2	90.2	6.8	35.1
IP-OD1	33 ± 7	−0.6 ± 1.8	13 ± 3	92.9	12.6	36.9
IP-OD2	34 ± 9	−1.6 ± 1.4	14 ± 3	96.5	14.5	35.4
IM1	22 ± 1	−0.5 ± 0.4	8 ± 1	93.6	8.1	48.4
IM2	22 ± 2	0.6 ± 1.1	11 ± 2	87.0	10.6	48.4

[a]Color difference referred to fresh sample.

sociated with the osmotic dehydration process and greater ones to VI. Despite the high ΔE values obtained in these cases, no significant changes of color hue and chrome attributes were observed, the main differences being due to loss of clarity in line with transparency gain.

CRYOPRESERVATION EFFECTS OF VI-OD

CRYOSTABILIZATION TECHNOLOGY

Cryostabilization technology emerged from food polymer science research and developed from a fundamental understanding of the critical physico-chemical and thermomechanical structure-property relationships, that underlie the behavior of water in all nonequilibrium food systems at subzero temperatures. Cryostabilization provides a means of protecting products, stored for long periods at typical freezer temperatures (e.g., $T = -18°C$) from changes in texture (e.g., "grain growth" of ice, solute crystallization), structure (e.g., collapse or shrinkage), and chemical composition (e.g., enzymatic activity, oxidative reactions such as fat rancidity, and flavor/color degradation). Such changes are exacerbated in many typical natural or fabricated foods whose formulas are dominated by low molecular weight (MW) carbohydrates (Slade and Levine, 1995).

The key to cryoprotection lies in controlling the physical state of the freeze-concentrated amorphous matrix surrounding the ice crystals in a frozen system, where deteriorative reactions mainly occur. This control can be exerted by controlling the physicochemical and mechanical properties of this matrix. These properties, as well as the diffusional process (Oliveira and Silva, 1992), change greatly from glassy to rubbery or to a liquid state when T_g' or T_m (ice melting temperature) are attained, respectively. The T_g' value of the product has a special technological significance. This is the glass transition temperature (T_g) of the maximally cryoconcentrated solution in the food, which is below the eutectic temperature (T_e) of the system. The range $T_g' - T_m$ corresponds to a rubbery state area for the amorphous concentrated matrix, where kinetics of physical and chemical changes are controlled by the William, Landel, and Ferry (WLF) equation (Williams et al., 1955) as concerns the temperature effect. In this model, the rate of reaction is greater when the difference $(T - T_g')$ increases. In this region, the high concentration of components leads to great rates of reactions, and so the deteriorative process occurs faster. Also, several studies showed that recrystalization temperature (or temperature where ice crystals begin to form) is very near to T_g'. The range $T_g' - T_m$ was also the region where growth of crystals occurs as a consequence of the increase in molecular mobility in the amorphous phase (Levine and Slade, 1988). However, at storing temperatures below T_g', small discreet ice crystals are embedded and immobilized in a continuous glassy matrix (Levine and Slade, 1988).

There are two possibilities to achieve an adequate food cryoprotection (Slade and Levine, 1995). One is the formulation of food with appropriate ingredients to elevate T_g' relative to T_f (freezer temperature), enhancing the product stability (Levine and Slade, 1989). Another possibility is the reduction of the water content of the product below W_g' (content of unfrozen water in the frozen product), allowing its complete vitrification, by means, for example, of osmotic dehydration with cryoprotectant solutions (Pinnavaia et al., 1988; Lazarides and Mavroudis, 1995). The typical freezer temperatures $(-18, -20°C)$ are normally greater than the T_g' of many products (Roos, 1995), especially when low MW carbohydrates are the main components, as is the case of fruits. So, deteriorative reactions and the growth of ice (and solute) crystals lead to a poor-quality frozen commodity (Blanshard and Franks, 1987). In fact, the cellular structure is practically totally destroyed in the thawed fruits, and important losses of flavor and color occur because of chemical reactions. Storage of frozen fruits at temperatures below T_g' could improve the quality of the thawed product a great deal, but this requires low temperatures in the freezer $(T_f < -40, -50°C)$ and, on the other hand, a small break off in the product storage temperature condition could also lead to a fast deterioration.

T_g' is a function of MW for both homologous and quasi-homologous families of water-compatible monomers, oligomers, and high polymers. The selection and use of appropriate ingredients in a fabricated product have allowed the food technologist to manipulate its T_g' and thus deliberately formulate the raising of T_g' relative to T_f, thus enhancing the product stability (Slade and Levine, 1991). Nevertheless, changes in composition of biological products such as fruits offer serious difficulties, although some modifications were described (Franks et al., 1977). Partial osmotic dehydration or immersion of the fruit in several solutions of some hydrocolloids were tried to preserve quality during frozen storage. Although some improvements in quality were described for dehydrofrozen fruits, no systematic studies, selecting appropriate cryostabilizers or cryoprotectants, were published. Nevertheless, extensive information about T_g' and W_g' (content of unfrozen water in the frozen product) of model solutions for both oligo and polysaccharides and mixtures has been reported (Levine and Slade, 1989; Slade, and Levine, 1991). In fact, there is a classification of these water-soluble compounds in cryostabilizers and cryoprotectants. Although some high MW carbohydrates had high T_g' and very low W_g' (cryostabilizers), low MW solutes have a characteristic combination of low T_g' and high W_g' that converts them into monomeric cryoprotectants for frozen stored products with a desirably soft-frozen texture but undesirably poor stability (Slade and Levine, 1991). The poor stability results from the relatively large ΔT $(T_f - T_g')$. The soft-frozen texture has a dual origin: the high W_g' value, which reflects the low ice content of the system, and the large ΔT value, which reflects the relatively large extent of softening of the non-ice portion of the system. In contrast, the use of cryoprotectants in biological products benefits only from the low ice content formed in cryopreserved specimens.

Complete vitrification is another possible approach to the prolonged cryopreservation of biological systems. This can be performed by ultrafast cooling rate during freezing and frozen storage below product T_g' or by reducing the product water content below W_g' by means, for example, of osmotic dehydration with cryoprotectant solutions (Andress, 1987; Tomasicchio et al., 1986; Pinnavaia et al., 1988). The former alternative is not practicable for foods because of its high cost. So, a high value of W_g' of the final system is important to preserve the wet (non-dried) characteristic of the treated product to a certain extent.

The VI operation permits cryoprotectants and cryostabilizers to be introduced deep inside the food's solid structure. Also, some decrease in water content by diffusional and osmotic mechanisms (associated with HDM in the vacuum osmotic dehydration) may be conducted at low temperatures with reasonable mass transfer rates, leading to a reduction in the amount of frozen water.

GLASS TRANSITION AND ICE-MELTING PROPERTIES

Table 7.4 shows the different effects of VI-osmotic dehydration treatments on apple samples, as a function of the different compositional changes shown in Table 7.1 and previously commented on. In Table 7.4, properties of ice-melting endotherms (T_g', T_m', ΔH) and freezable water content of the different samples can be observed. These properties were determined by differential scanning calorimetry (DSC 220CU SII, Seiko Instruments Inc., Chiba, Japan). To determine T_g' (midpoint), the incipient melting temperature (T_m') and ice-melting enthalpy samples were annealed for enough time to obtain coherent values and afterwards heated at 5°C/min.

TABLE 7.4. Properties of Ice-Melting Endotherm and Freezable
Water Content of Samples.

Sample	x_w (g/g)	T_g (°C)	T_m (°C)	ΔH (J/g)	x_{FW} (g_{FW}/g)	x_{NFW} (g_{NFW}/g)	W_g (g_w/g_{mcs})
F	0.85	−57.0	−39.0	275	0.81	0.02	0.21
OD1	0.72	−57.5	−40.0	192	0.64	0.14	0.22
OD2	0.61	−59.6	−40.9	152	0.50	0.16	0.22
IP	0.85	−57.1	−37.0	263	0.80	0.06	0.25
IP-OD1	0.76	−57.3	−41.7	190	0.68	0.19	0.25
IP-OD2	0.61	−57.5	−42.2	155	0.49	0.14	0.24
IM1	0.77	−56.3	−41.7	236	0.70	0.06	0.23
IM2	0.72	−56.0	−41.4	212	0.64	0.08	0.22

No significant differences between the T_g' values of the different samples were found, the average value being −57°C. That means that changes in solute composition due to treatments do not imply a notable modification in the behavior of apple native solutes. T_m' values of treated samples are also very close to that of the untreated one (−39°C). This temperature determines the storage temperature below which all freezable water takes the form of ice crystals if supercooling does not occur. At $T > T_m'$, a part of the freezable water is in a liquid state, thus contributing to the molecular mobility in the product, favoring its deterioration and ice crystals growing.

The values of freezable water of each sample (x_{FW}) were obtained from Equation (2). Both fresh samples and those VI with IP have a freezable water content significantly greater than dehydrated ones. This will imply a greater ice formation during a freezing process and, presumably, greater damage. In a_w reduced samples, as expected, the greater the a_w reduction, the lower the x_{FW}. Nevertheless, at the same a_w level, samples OD and IP-OD show less ice formation than samples IM. This seems to indicate that VI with low MW solutes will not favor frozen stability of apple samples.

$$x_{FW} = \frac{\Delta H_{exp}}{\Delta H_w^0} \qquad (2)$$

where ΔH_w^0 is melting enthalpy of pure water.

From x_{FW} values, the concentration of maximally cryoconcentrated solution (mcs), W_g', can be calculated [Equation (3)]. Values appear in Table 7.4, and it can be assumed that, although samples impregnated with pectin showed slightly higher W_g' values, there are no significant changes promoted by treatments, the average value of W_g' being 23 g water/100 g mcs.

$$W_g' = \frac{x_w - x_{FW}}{1 - x_{FW}} \qquad (3)$$

PRESERVATION OF MECHANICAL PROPERTIES, STRUCTURE, AND COLOR

Changes in mechanical properties due to freezing-thawing processes, related to structural damage caused by ice crystals and cell alterations, and also color development as a consequence of the advance of some deteriorative reactions could serve as indicators of cryopreservation effects of VI-OD pretreatments. It could be expected that samples with a lower water content would show a smaller change in the mechanical response, coherent with the less damage caused by ice crystals due to the diminution of freezable water content. On the other hand, ice crystals produce membrane injuries that could cause the outflow of intracellular liquid, thus increasing the viscous character of the product.

All these aspects have been analyzed in fresh and pretreated samples described in Table 7.1, frozen at $-40°C$ in sealed plastic bags for 2 h and stored at $-18°C$ for 24 h. After freezing, thawing was performed at $4°C$ for 24 h. Afterward, mechanical and structural properties and color of samples were evaluated.

Figures 7.2 and 7.3 show the effect of frozen-thawing treatments on initial (E_i) and asymptotic modulus (E_a), relaxation rate (B), and the relative relaxation levels (A).

In all samples, an important decrease in the modulus function after freezing-thawing can be observed. Nevertheless, this effect was more important in non-pretreated samples. Compared with unfrozen samples, the mechanical modulus shows a great reduction in all cases, which indicates a serious sam-

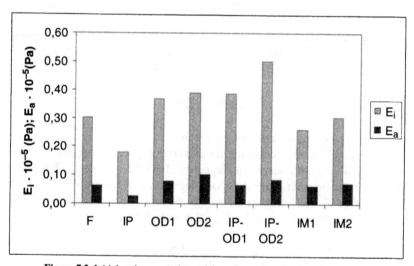

Figure 7.2 Initial and asymptotic modules of samples after freezing-thawing.

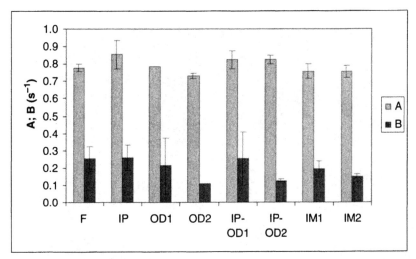

Figure 7.3 Relative relaxation level (*A*) and relaxation rate (*B*) of samples after freezing.

ple injury caused by freezing. Nevertheless, some differences in the modulus values were observed among the different thawed samples. Samples with the lowest water content, especially those VI with IPS, showed the highest values of modulus function, so they seem better preserved in their mechanical response. Stress relaxation level and relaxation rate does not show significant differences among the different frozen-thawed samples.

Structural damage of freezing-thawing treatments was also observed by cryo-SEM technique, performed by using a JEOL JSM-5410 microscope. Rectangular pieces 4 mm long, 1.5 mm wide, and 5 mm high were sliced from the center of each sample, frozen by immersion in Slush Nitrogen ($-210°C$) and after that, they were freeze-fractured, etched (at $-94.5°C$, 10^{-5} torr vacuum, for 15 min), gold coated, and viewed in the cold-stage SEM. After freezing treatments, the aspect of the samples (Figure 7.4) changed considerably compared with micrographs of samples before freezing.

In non-pretreated samples, the cell walls appear deformed (non-turgid), and the intercellular spaces appear flooded with intracellular liquid due to the rupture of cell membranes. This also occurred in OD samples. In some cases (especially in those VI with GM) vesiculations of disrupted plasmalemma can be observed. Impregnation with pectin combined with osmodehydration causes the cell walls to fold, but cell volumes remain quite well preserved. After freezing, this sample showed the least freezing injuries, this being coherent with the mechanical response observed. The viscous impregnated solution probably favors water vitrification, thus limiting ice and cryoconcentration damages.

Figures 7.5, 7.6, and 7.7 show color coordinates (*L**, hue, and chroma) of samples before freezing (BF) and 24 h frozen and thawed (AF). After freezing-

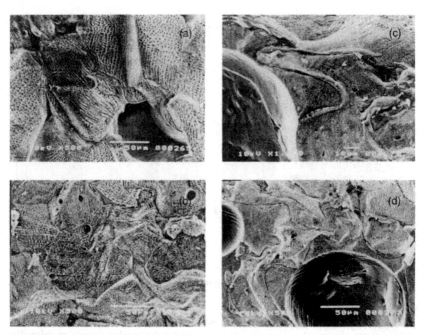

Figure 7.4 Cryo-SEM of frozen-thawed apple samples submitted to different pretreatments (a): Non-pretreated. (b): VI with GM of 61°Brix (OD2). (c) and (d): Submitted to osmotic dehydration (OD2); p: plasmalemma; cw: cell wall; v: vesicles; is: intercellular spaces.

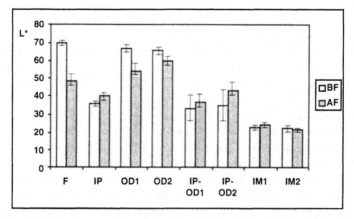

Figure 7.5 L^* values of apple samples before (BF) and after (AF) freezing-thawing.

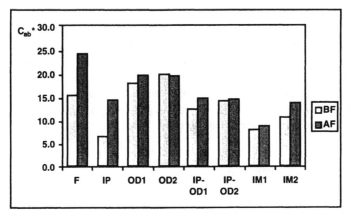

Figure 7.6 C_{ab}^* values of apple samples before (BF) and after (AF) freezing-thawing.

thawing, L^* decreases in non-impregnated samples compared with samples before freezing, but no significant L^* changes are observed in impregnated samples. This behavior can be explained because of the outflow of cellular liquid to the intercellular spaces, and the subsequent transparency gain, associated with the cell compartmentation loss during freezing. Frozen storage and thawing will imply apple browning mainly because of cellular breakdown and enzyme activity. In apple, browning will lead to an L^* decrease, hue shift to red wavelength (h_{ab}^* decrease), and chrome decrease. The observed changes in sample color coordinates show that browning was more intense in samples with the higher a_w. The best preserved samples in total color changes were the samples impregnated with IP and a_w depressed to 0.955. The limited diffusion of oxygen in the high viscous liquid phase of the sample could be the responsible factor.

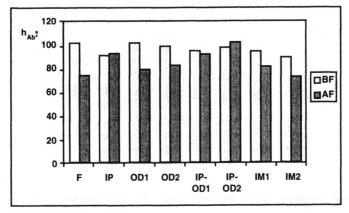

Figure 7.7 h_{ab}^* values of apple samples before (BF) and after (AF) freezing-thawing.

CONCLUSIONS

VI with polymers such as pectin, combined with water activity reduction by osmotic treatments, seems to improve the resistance of apple samples to freezing injuries in color and texture preservation. Impregnation with low MW solutes did not enhance frozen stability probably because of the introduction of a greater amount of water in the product pores without limiting the ice crystal growth, because of their low viscosity.

ACKNOWLEDGEMENTS

We thank the Comision Interministerial de Ciencia y Tecnología, the U.E (STD3 programme), and the Ciencia y Tecnología para el desarrollo (CYTED) Program for their financial support.

REFERENCES

Alzamora, S. M. and Gerschenson, L. N. 1997. Effect of water activity depression on textural characteristics of minimally processed fruits. In *New Frontiers in Food Engineering. Proc of the 5th Conf of Food Eng.* G. V. Barbosa-Cánovas, S. Lombardo, G. Narsimhan, and M. R. Okos, eds. New York: AICHE, pp. 72–75.

Andress, E. L. 1987. Studies of frozen and canned apple slices pretreated with vacuum hydration and hydrocolloids. *Dissertation Abstracts International* 48:931–932.

Barat, J. M., Chiralt, A., Albors, A., and Fito, P. 1999. Equilibration of apple tissue in osmotic dehydration. Microestructural changes. *Drying Tech.* 17:1375–1386.

Blanshard, J. M. V. and Franks, F. 1987. Ice crystallization and its control in frozen foods systems. In *Food Structure and Behaviour.* J. M. V. Blanshard and P. Lillford, eds. London: Academic Press, pp. 51–65.

Chiralt, A., Fito, P., Andrés, A., Barat, J. M., Martínez-Monzó, J., and Martínez-Navarrete, N. 1999. Vacuum impregnation: a tool in minimally processing of foods. In *Processing of Foods: Quality Optimization and Process Assessment,* F. A. R. Oliveira, J. C. Oliveira, eds. Boca Ratón, FL: CRC Press, pp. 341–356.

Crowe, J. H., Clegg, J. S., and Crowe, L. M. 1998. Anhydrobiosis: the water replacement hypothesis. In *The properties of water in foods IOSOPOW 6,* D. S. Reid, ed. London: Blackie Academic & Professional, pp. 440–453.

del Rio, M. A. and Miller, M. V. 1979. Effect of pretreatment on the quality of frozen melon balls. *Bull. de l'Institute Int. du Froid* 59:1172–1175.

Eskin, N. A. M. 1989. *Quality Preservations of Vegetables.* Boca Raton, FL: CRC Press.

Fito, P. 1994. Modelling of vacuum osmotic dehydration of food. *J. Food Eng.* 22:313–328.

Fito, P. and Chiralt, A. 1997. Osmotic dehydration: An approach to the modelling of solid food-liquid operations. In *Food Engineering 2000,* P. Fito, E. Ortega-Rodríguez, and G. V. Barbosa-Cánovas, eds. New York: Chapman & Hall, pp. 231–252.

Fito, P. and Chiralt, A. 1995. An update on vacuum osmotic dehydration. In *Food Preservation by Moisture Control: Fundamentals and Applications*, G. V. Barbosa-Cánovas, J. and Welti-Chanes, eds. Lancaster: Technomic Publishing Co., Inc., pp. 351–372.

Fito, P., Chiralt, A., Barat, J. M., and Martínez-Monzó, J. 2000. Vacuum impregnation in fruit processing. In *Trends in Food Engineering*, J. E. Lozano, C. Añón, E. Parada-Arias, and G. V. Barbosa-Cánovas, eds. Lancaster: Technomic Publishing Co., Inc., pp. 149–164.

Forni, E., Torreggiani, D., Crivelli, G., Mastrelli, A., Bertolo, G., and Santelli, F. 1987. Influence of osmosis time on the quality of dehydrofrozen kiwifruit. *Acta Hort* 282:425–433.

Franks, F., Asquith, M. H., Hammond, C. C., Skaer, H. B., and Echlin, P. 1977. Polimeric cryoprotectants in the preservation of biological ultrastructure I. *J. Microsc.* 110:223–238.

Giangiacomo, R., Torreggiani, D., Erba, M. L., and Messina, G. 1994. Use of osmodehydrofrozen fruit cubes in yogurt. *Ital. J. Food Sci.* 6(3):345–350.

Hawkes, J. and Flink, J. M. 1978. Osmotic concentration of fruit slices prior to freeze dehydration. *J. Food Proc. Pres.* 2(4):265–284.

Instituto Internacional del Frío. 1990. *Alimentos congelados. Procesado y distribución*, 3rd Ed. Zaragoza (España): Acribia, S.A.

Lazarides, H. N. and Mavroudis, N. (1995). Freeze/thaw effect on mass transfer rates during osmotic dehydration. *J. Food Sci.* 60(4):826–828, 857.

Levine, H. and Slade, L. 1988. Principles of cryostabilization technology from structure/property relationships of water-soluble food carbohydrates. A review. *Cryo-Lett.* 9:21.

Levine, H. and Slade, L. 1989. A food polymer science approach to the practice of cryostabilization technology: Comments. *Agric. Food Chem.* 1:315.

Martínez-Monzó, J., Martínez-Navarrete, N., Chiralt, A., and Fito, P. 1998a. "Mechanical and structural changes in apple (var. Granny Smith) due to vacuum impregnation with cryoprotectants," *J. Food Sci.*, 63(3):499–502.

Martínez-Monzó, J., Martínez-Navarrete, N., Chiralt, A., and Fito, P. 1998b. Osmotic dehydration of apple as affected by vacuum impregnation with HM pectin. In *Drying '98, Vol. A*, C. B. Akritidis, D. Marinos-Kouris, G. D. Saravacos, eds. Thessaloniki: Ziti Editions, pp. 836–843.

Martínez-Monzó, J., Martínez-Navarrete, N., Fito, P., and Chiralt, A. 1996. "Cambios en las propiedades viscoelásticas de manzana (Granny Smith) por tratamientos de impregnación a vacío. In *Equipos y Procesos para la industria de alimentos*. E. Ortega, E. Parada, and P. Fito, eds. Valencia: Servicio de Publicaciones de la Universidad Politécnica de Valencia, pp. 234–243.

Oliveira, U. and Silva, C. 1992. The influence of freezing on the diffusion of reducing sugars in carrot cortex tissues. *J. Food Sejence* 57(4):932–934.

Peleg, M. 1980. Linearization of relaxation and creep curves of solid biological materials. *J. Rheol.* 24:451–463.

Peleg, M. and Pollak, N. 1982. The problem of equilibrium conditions in stress relaxation analyses of solid food. *J. Texture Stud.* 13:1–11.

Pinnavaia, G., Dalla Rosa, N. I., and Lerici, C. R. 1988. Dehydrofreezing of fruit using direct osmosis as concentration process. *Acta Aliment. Polonica* 14:51–57.

Pitt, R. E. 1992. Viscoelastic properties of fruits and vegetables. In *Viscoelastic Properties of Foods*, M. A. Rao and J. F. Steffe, eds. London: Elsevier Applied Science, pp. 49–76.

Rao, V. N. M. 1992. Classification, description and measurement of viscoelastic properties of solid foods. In *Viscoelastic Properties of Foods*, M. A. Rao and J. F. Steffe, eds. London and New York: Elsevier Applied Science, pp. 3–47.

Reinbold, R. S., Hansen, C. L., Gale, C. M., and Ernstrom, C.A. 1993. Pressure and temperature during vacuum treatment of 290-kilogram stirred-curd cheddar cheese blocks. *J. Dairy Sci.* 76:909–913.

Robbers, M., Singh, R. P., and Cunha, L. M. 1997. Osmotic-convective dehydrofreezing process for drying kiwifruit. *J. Food Sci.* 62(5):1039–1047.

Roos, Y. H. 1993. Water activity and physical state effects on amorphous food stability. *J. Food Process. Preserv.* 16:433–447.

Roos, Y. H. 1995. *Phase Transition in Foods*. San Diego, New York, Boston, London, Sydney, Tokyo, Toronto: Academic Press.

Salvatori, D., Andrés, A., Chiralt, A., and Fito, P. 1998. The response of some properties of fruits to vacuum impregnation. *J. Food Proc. Eng.* 21:59–73.

Santerre, C. R., Cash, J. N., and Vannorman, D. J. 1988. Ascorbic acid/citric combinations in the processing of frozen apple slices. *J. Food Sci.* 53:1713–1716.

Shi, X. Q. and Fito, P. 1993. Vacuum osmotic dehydration of fruits. *Drying Technol.* 11:1429–1442.

Shi, X. Q., Fito, P., and Chiralt, A. 1995. Influence of vacuum treatment on mass transfer during osmotic dehydration of fruits. *Food Res Int.* 28:445–454.

Slade, L. and Levine, H. 1991. Beyond water activity: Recent advances based on an alternative approach to the assessment of food quality and safety. *Crit. Rev. Food Sci. Nutr.* 30(2–3):115–360.

Slade, L. and Levine, H. 1995. Glass transitions and water-food structure interactions. In *Advances in Food and Nutrition Research, Vol. 38*, J. E. Kinsella, ed. San Diego: Academic Press, pp. 103–269.

Tomasicchio, M., Andreotti, R., and Giorgi, A. 1986. Osmotic dehydration of fruits II. Pineapples, strawberries and plums. *Industria Conserve* 61:108–114.

Torregiani, D. 1995. Technological aspects of osmotic dehydration in foods. In *Food Preservation by Moisture Control: Fundamentals and Applications*, G. V. Barbosa-Cánovas and J. Welti-Chanes, eds. Lancaster: Technomic Publishing Co., Inc., pp. 281–304.

Torreggiani, D., Forni, E., Crivelli, G., Bertolo, G., and Mastrelli, A. 1987. Researches on dehydrofreezing of fruit. *Part 1: Influence of Dehydration Levels on the Product's Quality*, Proceedings of XVII Int. Congress of Refrigeration, Vienna, pp. 461–467.

Williams, M. L., Landel, R. F., and Ferry, J. D. 1955. Temperature dependence of relaxation mechanisms in amorphous polymers and other glass-forming liquids. *J. Am. Chem. Soc.* 77:3701–3707.

Zozulevich, B. and D'yachenco, E. N. 1969. Osmotic dehydration of fruits. *Konservn. Ovoshchesusch. Prom.* 7:32–34.

CHAPTER 8

Yield Increase in Osmotic Processes by Applying Vacuum Impregnation: Applications in Fruit Candying

J. M. BARAT
G. GONZÁLEZ-MARIÑO
A. CHIRALT
P. FITO

INTRODUCTION

O SMOTIC dehydration (OD) implies three main fluxes in the osmosed food. The uptake of solutes from the osmotic solution and the losses of water (usually the greater flux) and native solutes (qualitatively very important) (Torreggiani, 1993).

OD can be used in food industries to obtain different kind of products:

- Sugared products in which a sweetness characteristic would be desirable (e.g., correction of the sweet/sour relationship in fruits).
- Intermediate moisture products in which the combination of the water activity depression with other preservation factors would enable us to obtain a "fresh-like" product that could be consumed directly or used in food formulation (e.g., yogurt).
- Dried products previously osmotically dehydrated.
- Jams and marmalade in which the water removal from the product could be performed by osmosis (e.g., at low temperatures), and the osmotic solution reconcentration could be performed in a separate operation.
- Salted products (such as ham or cheese) in which the salt uptake occurs simultaneously with the water removal at low temperatures.
- Candied fruits and vegetables, which are highly sugar-enriched products, usually processed in an artisan way. In these, a great sugar gain is promoted by using sugar solutions of increasing concentrations and applying high temperatures during long treatments.

In these processes, osmotic dehydration is conducted separately or in combination with other operations, such as air drying, pasteurization, etc.

Different kinds of products imply a different ratio of water loss/external solid uptake, which must define either the operation conditions in OD or the conducting of some sample pretreatment. In the first case, the solute molecular weight, process pressure, and temperature or sample size must be controlled (Lazarides et al., 1995). For pretreatment, different actions can be performed: vacuum impregnation (VI) with external solution or blanching, which will promote solute gain, and coating with edible films, which limits the effect.

In this chapter, the influence of some of the above-mentioned factors on the yield of some of the most typical osmotic processes, e.g., candy fruits, will be discussed specifically.

MATERIALS AND METHODS

Two sets of OD experiments were performed with sucrose solutions at controlled temperature and pressure. Atmospheric pressure was maintained in OD, and a pulse of vacuum pressure (180 mbar) was applied in the vacuum osmotic dehydration (PVOD) experiments, for the first 5 min afterward carrying on at atmospheric pressure.

In the first set of experiments, short-time processes (far away from the concentration equilibrium) were accomplished. Apples (*Granny Smith*) were peeled, decored, and sliced perpendicular to the apex-base direction. Slices 10 mm thick, having 64 mm and 20 mm external and internal diameter, respectively, were cut. Apple slices were osmosed for 15, 30, 45, 60, 120, and 420 min in a Pilot Plant at a flow rate of solution, which ensured the internal control of mass transfer rate. Working temperatures were 30 and 40°C, and sucrose solution concentrations were 45, 55, and 65°Brix. In all cases, the solution-fruit ratio was great enough (50:1) to avoid significant changes in the solution concentration during the process.

In the second set of experiments, long-time processes (near the equilibrium situation) were assayed by using apple (*Granny Smith*), pear, mango, kiwi, and pineapple. Fruits were peeled, and cylindrical samples (2 cm long and 2 cm diameter) were cut with a borer, with the longitudinal axes in the fruit apex-base direction, except for pineapple in which 1-cm-thick slices were obtained. Working temperatures were 30 and 40°C, and sucrose solution concentrations were 45, 55, and 65°Brix. Fruit samples from the same fruit piece were placed into a flask that contained a non-stirred sucrose osmotic solution. The solution-fruit ratio was higher than 20 in all cases. Temperature of the system was maintained by placing the flask into an oven. Potassium sorbate (2000 ppm) was added to the osmotic solutions to stabilize the samples microbiologically. For each treatment, the weight and volume of three of the samples were controlled throughout the equilibration time.

ANALYTICAL DETERMINATIONS

Sample volume was determined by direct measure (slices) or with a solid pycnometer (cylinders) by using, in each case, the respective osmotic solution as reference liquid. Moisture content of samples was determined by placing the samples in a vacuum oven until constant weight was reached (AOAC 20013; A.O.A.C., 1980). Soluble solids were analyzed by measuring the refraction index in a refractometer (Atago, NAR T3, Japan) at 20°C. To obtain a clear refractometric measure, samples were homogenized with distilled water by using an ultraturrax mixer. Refractometric measures were corrected because of the effect of the dilution factor to obtain the original sample concentration of soluble solids in the food liquid phase (FLP).

RESULTS AND DISCUSSION

VI EFFECT ON THE YIELD OF SHORT-TIME FRUIT OSMOTIC DEHYDRATION PROCESSES

Working with apple slices at various process conditions, it has been seen experimentally that when VI is applied before the osmotic dehydration processing, product weight is heavier than that of those not impregnated (Figure 8.1).

There are some reasons that explain the heavier weight when VI is applied for short-time processes. All these reasons are based on porous gas replacement by the osmotic solution (OS).

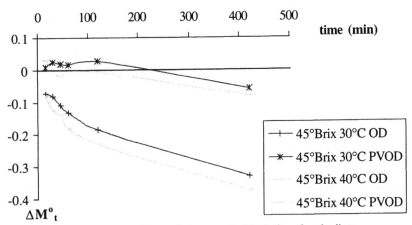

Figure 8.1 Weight changes during osmotic dehydration of apple slices.

(1) Gas replacement by osmotic solution implies a fast initial weight increase. Other authors (Fito et al., 1994; Fito et al., 1998) have observed the mentioned weight increase (see Figure 8.1). It appears even though during the vacuum pulse there is an important amount of water loss due to the sample dehydration while submerged in the osmotic solution (5 min of vacuum pressure and 10 min of atmospheric pressure).

(2) Vacuum impregnated samples need to lose less water to reach the equilibrium concentration value. The result of the initial OS gain is that in a very short time, the average FLP concentration increases, being closer to the OS concentration and so much nearer to the equilibrium concentration. When FLP concentration is closer to the OS concentration, the quantity of water that must be lost by samples is lower, implying a smaller weight decrease (Figure 8.1).

(3) Vacuum impregnation is more favorable to solute uptake than to water loss. Gas replacement by OS during vacuum impregnation increases the pathway for solute uptake and water loss (Fito, 1994; Fito and Pastor, 1994). These new pathways for mass transfer are more favorable to solute uptake (Fito, 1994; Barat et al., 1998a) because of the absence of cell membranes in those spaces.

It has been observed that vacuum impregnation reduces case hardening effect when working with high concentration osmotic solutions (Barat et al., 1998a). This would be a new effect favorable to solute uptake when working with PVOD instead of OD.

In the experiments, it can be easily seen that water loss increases for OD and solute uptake for PVOD. In Figure 8.2, the experimental results obtained when working with the 65°Brix OS are shown as an example of the above-mentioned behavior.

In conclusion to the observed behavior, the use of VI as a pretreatment to increase the process yield for short-term OS treatments is proposed. As can be seen in Figure 8.3, product weight is bigger using VI, for the same FLP concentration value. This is observed for all the experimental concentrations and temperatures. Obviously, for short-time processes (far away from the concentration equilibrium), when a determined value of FLP concentration or a_w are desired, the yield of the process would be greater if VI pretreatment would be used.

For results shown in Figure 8.3, it is important to note that the lower the OS concentration, the greater the process yield. This effect would be explained by taking into account the lower viscosity for low concentrated OSs. When the viscosity is lower, the transport of sugars seems to be favored, which leads to less transport of water. Another remarkable aspect is that for the observed range of temperatures, there are no appreciable differences in the behavior at 30 and 40°C, which would imply that temperature would affect the process kinetic (Figure 8.1) but not the process yield.

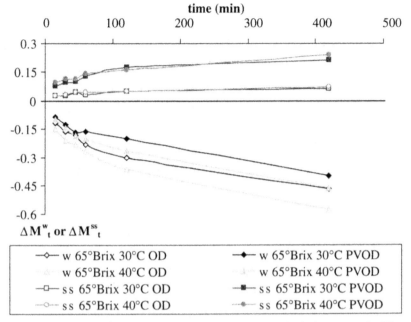

Figure 8.2 Water loss and solute gain throughout the osmotic dehydration process for OD and PVOD.

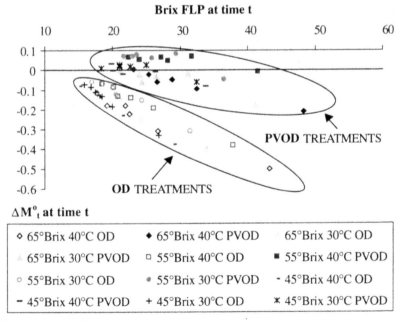

Figure 8.3 Relationship of ΔM_t^o with FLP concentration in osmotic dehydration of apple, using sucrose/water osmotic solution at 30 and 40°C.

For volume changes, it was observed (Figure 8.4), that for PVOD samples there was a bigger initial volume reduction due to the structure contraction when the atmospheric pressure was restored (Fito et al., 1996). Nevertheless, when the process advances, the volume reduction for OD samples becomes greater because of the greater amount of water lost.

VI EFFECT ON THE YIELD OF LONG-TIME FRUIT OSMOTIC DEHYDRATION PROCESSES

In a previous article (Barat et al., 1998b) the behavior of apple cylinders immersed in an OS was described until the equilibrium was reached. Three main steps were described:

- In the first one, the concentration gradients were predominant, and as a consequence of the big amount of water lost, there was a volume reduction until a minimum value was reached. This period was from the first moment until the concentration gradient disappeared.
- In the second period, when the concentration gradient disappeared, the shrunken sample, with a big amount of stress accumulated in its structure swelled, and there was a volume recovery by sucking in the OS. This period finished when the volume recovery stopped.
- In the third period, the remaining gas phase is substituted by the OS, until weight changes cease and the final equilibrium situation is reached.

The three stages described above are more or less important as a function of the product, OS concentration, pretreatment (e.g., VI), and temperature and coexist at the same time, depending on the sample point considered in each case.

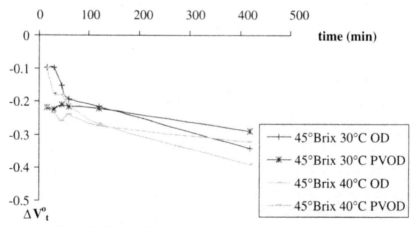

Figure 8.4 Volume changes during osmotic dehydration of apple slices.

In this point, the experimental results obtained for the same process conditions, but changing the sample shape (cylinders instead of slices), are shown. The processes where the end point is near the equilibrium are called "long-time" processes.

When the VI effect on weight evolution throughout the time was studied for long-time processes, the observed behavior was similar to that shown in the short-time studies. The use of VI implies a bigger process yield in all cases (Figure 8.5).

For long-time osmotic dehydration processes, a very important increase in fruit weight can be observed. This phenomenon was observed and explained by Barat et al. (1998b). There are two reasons that explain the increase in weight. First, the increase in volume due to the relaxation of the cell matrix, release the stress that appeared in the structure during its shrinkage in the first hours of the process. The other reason that explains the increase in weight is that the remaining gas in the porous structure is slowly replaced by the OS (Barat et al., 1998b).

For the volume recovery for long-time processes, it is observed that the degree of reversibility depends on the maximum volume loss, as can easily be seen in Figure 8.6. When a sample is vacuum impregnated, the maximum value of volume reduction is lower than that for samples not vacuum impregnated, and so the final volume recovered is greater (Figures 8.4 and 8.6). This would be the main reason that enables us to obtain better yields for long-time processes when using VI.

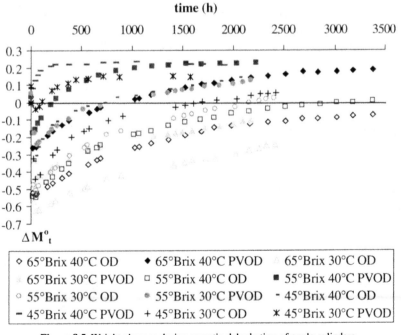

◇ 65°Brix 40°C OD	◆ 65°Brix 40°C PVOD	65°Brix 30°C OD
65°Brix 30°C PVOD	□ 55°Brix 40°C OD	■ 55°Brix 40°C PVOD
○ 55°Brix 30°C OD	◉ 55°Brix 30°C PVOD	- 45°Brix 40°C OD
- 45°Brix 40°C PVOD	+ 45°Brix 30°C OD	✕ 45°Brix 30°C PVOD

Figure 8.5 Weight changes during osmotic dehydration of apple cylinders.

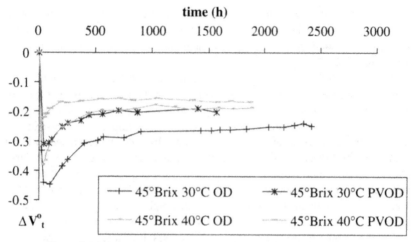

Figure 8.6 Volume changes during osmotic dehydration of apple cylinders.

Gas release from the porous structure throughout the process is very low and is affected by process temperatures. The absence of gas in fruit would lead to a translucent aspect, which is desired in the case of candied products. This is the reason why it is more difficult to obtain this kind of products working with OD at low temperatures (Barat et al., 1998c).

It has been observed that other fruits, such as pear, mango, kiwi, and pineapple, show a similar behavior during the osmotic processes than that previously observed for apple (Barat et al., 2000) (Figure 8.7). In all cases, vacuum impregnated samples at the beginning of the osmotic process have a better yield for long-time treatments than those not vacuum impregnated.

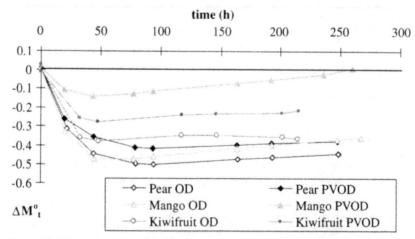

Figure 8.7 Changes in weight of osmotically dehydrated fruit cylinders at 30°C and 55°Brix OS concentration (from Barat et al., 2000. Reproduced with permission from Forum S.r.l.).

It can be seen that mass changes are lower in the case of PVOD than in the case of OD, as was observed with apple cylinders. The most significant effect of vacuum impregnation on the process yield was observed for mango (ϵ_e: 5.9), probably because of its bigger porosity compared with the pear (ϵ_e: 2.7) and kiwi (ϵ_e: 0.5) (Salvatori, 1997).

In Figure 8.8, the results obtained when working with pineapple at 40°C are shown, and the same behavior is observed as for the other fruits. The big differences of the OD and PVOD process could be explained by the relatively high porosity of this fruit (ϵ_e: 6.63) (González-Mariño, 1999).

The different solid matrix rigidity and structure, in addition to the porous volume fraction, could explain the different behavior observed among the fruits assayed. In this kind of process in which the equilibrium is reached, the structural behavior of the sample plays a very important role because of the solid matrix relaxation after the initial shrinkage promoted by the water loss (Barat et al., 1998b).

CANDYING OF PINEAPPLE

Up to this point, the observed results show the following aspects for long-time osmotic processes:

- The aim of combining increasing concentrations is to increase the process yield because the maximum volume loss can be reduced, and so the volume recovery would be less irreversible (bigger final volume).
- The use of vacuum impregnation in combination with low-process temperatures enables us to obtain translucent products (without residual gas) in a short time.

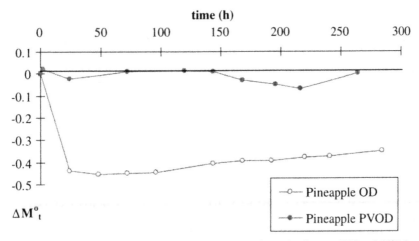

Figure 8.8 Changes in weight of osmotically dehydrated pineapple slices at 40°C and 55°Brix OS concentration.

• The use of vacuum impregnation can substantially increase the yield in long-time processes, such as the case of the candying process.

All these reasons make the use of VI in candying processes when working at low temperatures a reasonable possibility, as was proposed by Barat et al. (1998c).

In this point, results obtained in the candying process of pineapple are shown. Pineapple is a product consumed fresh and candied, and this is the reason why it was chosen in this candy process attempt.

In Figure 8.9, weight changes for long-time processes when working with pineapple slices (1-cm-thick) can be observed for OD and PVOD. A similar behavior to the other studied products and a smaller weight loss when low OS concentrations are used can be seen. It is important to note that even when 65°Brix are used directly in the osmotic dehydration of samples, the use of VI at the beginning reduces dramatically the maximum weight decrease, being closer to the 45°Brix treatment than when VI is not applied.

In Figure 8.10, weight changes for pineapple can be observed when a combination of concentrations similar to the industrial process is used. A marked effect of this concentration combination and the use of VI on process yield is observed. The increase in process yield due to VI is approximately 45%, which would be very interesting industrially (Barat et al., 1998d). Final product density is another important property in this assay, due to its relationship with the translucent aspect of the product and to its relevance to handling aspects (storage, transport, etc.), which would be more beneficial in the case of the vacuum impregnated product.

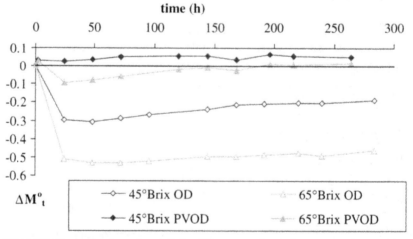

Figure 8.9 Changes in pineapple weight for different osmotic solution concentrations working at 40°C.

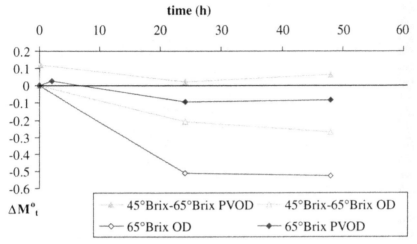

Figure 8.10 Pineapple weight changes at 40°C and osmotically dehydrated for 48 h using 65°Brix and by combining 45 and 65°Brix osmotic solutions.

The final product density is bigger for PVOD, and the porosity value is calculated as near 0, whereas for OD it is bigger. So, it is also concluded that if a low-temperature process is desirable, VI is needed to give it a translucent aspect.

CONCLUSIONS

The use of vacuum impregnation in osmotic processes, when a solute-enriched product is desired, implies much bigger yields than when VI is not applied. For short-time processes, the increase in solute gain and decrease in water loss are the combined factors that imply a bigger process yield. For long-time processes, the combination of the previous factors with a lower maximum shrinkage implies a bigger yield when VI is applied. A commercial product from pineapple was obtained by using a long-time process (candy process), and it was observed that the use of VI implied an increase in process yield much bigger than when atmospheric pressures were used throughout the whole process.

NOMENCLATURE

$\Delta M_t^{ss} = \dfrac{M_t^{ss} - M_0^{ss}}{M_0^o}$: soluble solids weight change at time t (g/g)

$\Delta M_t^o = \dfrac{M_t^o - M_0^o}{M_0^o}$: total weight change at time t (g/g)

$\Delta V_t^o = \dfrac{V_t^o - V_0^o}{V_0^o}$: volume change at

time t (cm^3/cm^3)

$\Delta M_t^w = \dfrac{M_t^w - M_0^w}{M_0^o}$: water weight

change at time t (g/g)

a_w: water activity

FLP: food liquid phase

M: sample mass (g)

OD: osmotic dehydration at atmospheric pressure

OS: osmotic solution

PVOD: pulsed vacuum osmotic dehydration

ss: soluble solids

t: time (h)

v: volume (cm^3)

VI: vacuum impregnation

w: water

Superscripts:

o: total mass or volume

ss: soluble solids

w: water

Subscripts:

t: values at time t

REFERENCES

AOAC. (1980). *Official Methods of Analysis.* Association of Official Analytical Chemists, Washington D.C.

Barat, J. M., Chiralt, A., and Fito, P. (1998a). Case hardening in osmotic dehydration of fruits. Effect of vacuum impregnation. In V. Gaukel y, W. E. L. Spiess, eds. *3rd Karlsruhe Symposium: European Research towards Safer and Better Food. Proceedings Part 2.* Druckerei Grässer, Karlsruhe, Germany. pp. 363–370.

Barat, J. M., Chiralt, A., and Fito, P. (1998b). Equilibrium in cellular food osmotic solution systems as related to structure. *J. Food Sci.* 63(5):836–840.

Barat, J. M., Lloría, R., Chiralt, A., and Fito, P. (1998c). Vacuum impregnation: A useful tool in candied fruit/vegetables processing. In V. Gaukel y, W. E. L. Spiess, eds. *3rd Karlsruhe Symposium: European Research towards Safer and Better Food. Proceedings Part 2.* Druckerei Grässer, Karlsruhe, Germany, pp. 371–376.

Barat, J. M., Lloría, R., Chiralt, A., and Fito, P. (1998d). Mejoras en el objeto de la patente principal n° P9300805 por procedimiento de flujo alternado para favorecer los intercambios líquidos de productos alimenticios y equipo para realizarlo (con aplicación específica a la fabricación de fruta confitada). P9800733.

Barat, J. M., Lloría, R., Chiralt, A., and Fito, P. (2000). Use of vacuum impregnation in fruit candying at low temperatures. In: Marco Dalla Rosa, ed. *Industrial Applications in Food Processing.* Forum, Udine, Italia, pp. 49–55.

Fito, P. (1994). Modeling of vacuum osmotic dehydration of food. *J. Food Eng.* 22:313–328.

Fito, P. and Pastor, R. (1994). On some non-diffusional mechanims occuring during vacuum osmotic dehydration. *J. Food Eng.* 21:513–519.

Fito, P., Andrés, A., Chiralt, A., and Pardo, P. (1996). Coupling of hydrodynamic mechanism and deformation-relaxation phenomena during vacuum treatments in solid porous food-liquid systems. *J. Food Eng.* 27:229–240.

Fito, P., Andrés, A., Pastor, R., and Chiralt, A. (1994). Vacuum osmotic dehydration of fruits. In: R. P. Singh and F. A. R. Oliveira, eds. *Minimal Processing of Foods and Process Optimization. An interface.* Boca Raton, CRC Press, pp. 117–121.

Fito, P., Chiralt, A., Barat, J., Salvatori, D., and Andrés, A. (1998). Some advances in osmotic dehydration of fruits. *Food Sci. Technol. Int.* 4(5):329–338.

Gonzáles-Mariño, G. (1999). Viabilidad de la piña colombiana var. Cayena lisa para su industrialización combinando las operaciones de Impregnación a Vacío, Deshidratación Osmótica y secado por aire caliente. PhD thesis, Valencia, Spain.

Lazarides, H. N., Katsanides, E., and Nicolaides, A. (1995). Mass transfer kinetics during osmotic preconcentration aiming at minimal solid uptake. *J. Food Eng.* 25(2):151–166.

Salvatori, D. (1997). Deshidratación Osmótica de Frutas: Cambios Composicionales y Estructurales a Temperaturas Moderadas. PhD. thesis, UPV, Valencia, España.

Torreggiani, D. (1993). Osmotic dehydration in fruit and vegetable processing. *Food Res. Int.* 26:59–68.

CHAPTER 9

Orange Peel Products Obtained by Osmotic Dehydration

M. CHÁFER
M. D. ORTOLÁ
A. CHIRALT
P. FITO

INTRODUCTION

A fast growth of the citrus industry in the past 25 years has occurred in line with the sharp advance of agricultural science and technology. The citrus world production in 1980 was 56.61 million metric tonnes, whereas in 1990 it was 67.63 million metric tonnes (FAO). Citrus fruit is processed mainly into juice, as frozen concentrates, pectin, peel, and seed oil and squash. Food items produced from peel are brined and candied peels, marmalades, syrups, and peel products for food seasoning. The peel juice, or press liquor, can also be used as a fermentable carbohydrate source for the production of feed yeast, industrial alcohol, vinegar, butylene, and lactic acid (Cohn and Cohn, 1996).

Citrus peel by-products may be a source of additional revenue for citrus processors, and new product applications should be investigated. In this sense, the nutritional and health properties of some peel components such as pectin, flavonoids, carotenoids, or limonene make the research more interesting (Girard and Mazza, 1998; Braddock, 1999). Among the health benefits, the effects of pectin on glycemic control, serum cholesterol concentration, cancer prevention, and control of mineral balance stand out (Baker, 1980; Borroto et al., 1995; Larrauri et al., 1995). Likewise, the effects of limonene on cancer prevention (Chander et al., 1994; Philips et al., 1995) and the vitamin activity of carotenoids (provitamin A) (Cerezal and Piñera, 1996) have been investigated.

Citrus peel consists of two parts: the albedo portion, which is the inner white spongy layer, and the flavedo or external colored part. Albedo is the main pectin source and consists of large parenchymatous cells with a very porous

93

structure (Spiegel-Roy and Goldshimdt, 1996). The flavedo part contains carotenoids (main pigments in citrus fruits) and the oil glands, which are the raised structures in the skin with the essential oils whose major component is limonene.

Recent market studies show an increasing demand for production and making of organic, fresh-like foods (MAPA, 1991; Wirthgen, 1994). Therefore, the development of new processing methods that are more economic and suited to consumer necessities must be conducted to improve food nutritional quality, minimizing thermal treatments and the addition of artificial additives. These processes must be environmentally aware, using closed cycles that reduce residues. The use of citrus peel (by-product from citrus juice industry) in food industry will permit the reduction of contaminants while supplying interesting products for human consumption.

In this sense, osmotic dehydration (OD) represents a good choice to develop high-quality orange peel products, taking advantage of its interesting composition, increasing its sweetness, and improving its sensory acceptability. In OD process, water is transferred from the fruit to a concentrated sugar or salt external solution without undergoing a phase change, thus not needing intensive heat treatment. The use of fruit concentrates, such as concentrated grape must as osmotic solutions, allows us to design "whole fruit products" greatly appreciated by consumers. Osmotic dehydrated fruits have a good flavor, aroma, and nutritional value because of the scarce reduction in the mineral and vitamin contents (Ponting, 1973; Yang et al., 1978; Vial et al., 1991).

Recently, OD by applying vacuum pressure has been extensively studied (Fito, 1994; Fito and Chiralt, 1997). Vacuum conditions promote mass transfer kinetics because of the action of hydrodynamic mechanisms (HDM), coupled with diffusional phenomena. Vacuum is specially effective in promoting HDM in highly porous products when it is applied in a tank containing the product immersed in a liquid phase. The internal gas in the pores is expanded and flows out, whereas external liquid penetrates because of capillary forces. When atmospheric pressure is restored, the residual gas is compressed, leading to a practically complete impregnation of the product pores by the external solution. This phenomenon implies a fast compositional change in the product at the same time as that product modifies its mass transfer behavior because of its porosity reduction (Fito, 1994; Fito and Pastor, 1994).

In this work, some aspects of OD of orange and mandarin peels are discussed as an alternative to develop new minimally processed citrus peel products, to be used as food ingredients in the manufacture of dairy products, confectioneries or pastries, marmalades (Shi et al., 1996), etc. Special emphasis is placed on the effect of vacuum pulse on OD effectiveness and product final properties.

FEASIBILITY OF VACUUM IMPREGNATION WITH AN EXTERNAL SOLUTION

In vacuum impregnation (VI) of a porous product, the exchange of the internal gas or liquid in the pores for an external solution with specific/selected composition occurs. This is usually conducted by applying vacuum pressure (P_1) (in the order of 50 mbar) in a tank containing the product immersed in the solution for a short time (t_1). Afterward, atmospheric pressure (P_2) is restored for a time (t_2), thus leading to the liquid flowing in. The possibility of introducing an external solution with specific/selected solutes into the product pores made the VI a tool in the fruit processing. Impregnation of the fruit pores has been seen to occur without vacuum action when the fruit remains immersed in a liquid phase for a long time (e.g., syrup canned and candied fruits) because of the capillary forces and pressure and temperature fluctuations in the system (Barat et al., 1998). The behavior of fruit tissue throughout VI will be greatly affected by product porosity and pore morphology, as well as by the viscoelastic properties of solid matrix, because pressure change could promote sample volume changes and porosity variations. The flow properties of the external solution also affect the fruit VI response (Chiralt et al., 1999). The response of several fruits and vegetables to VI has been studied by using isotonic solutions to limit mass transfer phenomena other than that promoted by HDM, applying a gravimetric methodology previously described (Fito et al., 1996; Salvatori, 1998).

Response of orange and mandarin peel to VI is reflected in Table 9.1 in characteristic parameters: sample volume deformation (γ) and impregnation (X) at each process step: at the end of vacuum period (γ_1 and X_1) and after restoring the atmospheric pressure (γ and X). Values of γ and X are expressed as volume fractions of the initial product. Sample effective porosity (ϵ_e) was determined from the values of VI parameters applying Equation (1), according to the VI reported model where ($\equiv P_2/P_1$) is the compression ratio (Fito et al., 1996). Density of orange and mandarin peels is very low, the values being 774 and 849 kg/m^3, respectively, because of the very high porosity of albedo. Peel water content is in the order of 75% and soluble solid contents of 15%. So, insoluble solids (product solid matrix) constitute about 10% (w/w)

TABLE 9.1. Response of Mandarin and Orange Peel to VI with Isotonic Solutions.

Citrus	γ	X	ε_e	ε_T
Orange peel var. *Valencia Late*	0.14 ± 0.06	0.37 ± 0.06	0.23 ± 0.04	0.29 ± 0.02
Mandarin peel var. *Satsuma*	0.21 ± 0.12	0.41 ± 0.07	0.14 ± 0.07	0.227 ± 0.004

of the product, containing 90% (w/w) of liquid phase (water plus solutes), with a water activity (a_w) of 0.97.

From Table 9.1 the great sensitivity of peels to VI can be observed as expected from their highly porous structure. High values of X and γ were reached at the end of the process when atmospheric pressure was restored in the second step of VI. Approximately 37–41% of sample initial volume may be impregnated with the isotonic solution by effect of HDM, in line with a sample swelling of 14% and 21% for orange and mandarin, respectively. By applying Equation (2) (Fito et al., 1996), the product effective porosity, referred to the sample initial volume, was estimated to be 23% and 14% in orange and mandarin peel, respectively; these values were in the range of those obtained for highly porous fruits such as apples (Salvatori et al., 1998). Porosity obtained by volume displacement in a pycnometer using isotonic solution as reference liquid is in the same order.

$$\epsilon_e = \frac{(X - \gamma)r + \gamma_1}{r - 1} \tag{1}$$

The response of citrus peel to VI seems similar to that of a sponge coherent with its described morphology, with great intercellular spaces and nonturgid cells (Spiegel-Roy and Goldschimdt, 1996). In Figure 9.1(a), micrographs of fresh orange peel obtained by cryo-SEM observations show the big void spaces in the albedo zone. The enlarged shape of the cells and the ruptures of cell junctions during fruit ripening can also be observed (Storey and Treeby, 1994). The microstructure of vacuum impregnated sample with isotonic solution [Figure 9.1(b)] shows the presence of the external solution filling the sample voids. This is reflected through the continuous dentritic aspect of the sample where cells appear occluded. This aspect is generated during the edging step in the microscope because of sample ice sublimation and cryoconcentration of the residual solution (Bomben and King, 1982).

The particular behavior of orange peel in VI process makes it a vehicle (or matrix) for incorporating physiologically active compounds, by VI with their solutions, to obtain a functional food. In this sense, the health properties of peel itself may be reinforced or fortified with other interesting compounds. Figure 9.2 shows the predicted values of mass fraction (x_i^I) of an active component (i) impregnated by VI in the peel as a function of its mass fraction in the external solution (y_i). Equations (2) and (3), obtained from mass balances in the VI product, are used in the predictive model (Chiralt et al., 1999). Even at low concentration of i in the external solution, a considerable amount can be introduced in the peel because of its high porosity.

$$x_i^I = \frac{x_i^0 + x_{HDM}y_i}{1 + x_{HDM}} \tag{2}$$

$$x_{HDM} = X \frac{\rho^{IS}}{\rho^0} \tag{3}$$

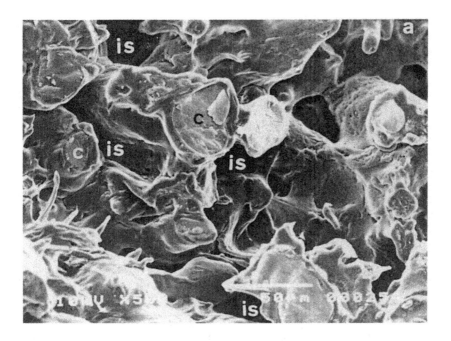

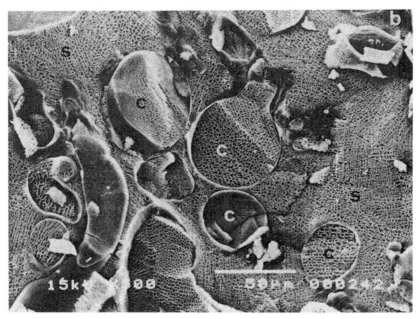

Figure 9.1 Cryo-SEM micrographs of albedo zone in fresh mandarin peel (a) and in vacuum-impregnated mandarin peel with an isotonic solution (b) (×500); c: cell; is: intercellular space; s: isotonic solution in the intercellular spaces.

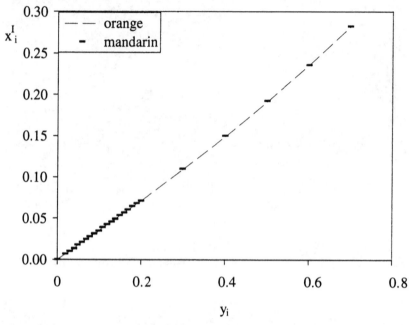

Figure 9.2 Mass fraction of a component *i* reached in the impregnated orange or mandarin peel, as a function of concentration in the impregnating solution (y_i).

BEHAVIOR OF ORANGE PEEL IN OD PROCESSES

Vacuum treatments lead to faster osmotic processes due to the coupled action of HDM with pseudo-diffusional mechanisms (PDM). VI of the sample during OD takes place by applying a vacuum pulse (for approximately 5 min) in the tank at the beginning of the process [pulsed vacuum osmotic dehydration (PVOD)]. This implies a fast compositional change in the product that will affect the osmotic driving forces and mass transfer kinetics (Barat et al., 1997; Fito and Chiralt, 1997). VI also implies changes in the samples in terms of mass, volume, density, and structure (Barat et al., 1998).

According to Fito and Chiralt (1997), the kinetic analysis of sample OD has been approached from two points of view related with process yield and product stability predictions, respectively. The weight change of the product, in water and solute gains [Equations (4) and (6)], was modeled for different treatments to evaluate the process yields. The compositional change of the fruit liquid phase defines the product stability and final quality and was also evaluated and modeled.

$$\Delta M_w = (M_t x_w^t - M_0 x_w^0)/M_0 \qquad (4)$$

$$\Delta M_s = (M_t x_s^t - M_0 x_s^0)/M_0 \qquad (5)$$

$$\Delta M_t = \Delta M_s + \Delta M_w \qquad (6)$$

KINETICS OF WATER AND SOLUTE GAINS

In previous works (Chafer et al., 1998; Chafer et al., 2000a; 2000b), linear relationships were found between the solute (ΔM_s) and water gains (ΔM_w) defined by Equations (4) and (5) and the square root of time. So, kinetic Equations (6) and (7) (Fito and Chiralt, 1997) were used for modeling. In these equations, the constant k_{i0} represents a component (water or solute) gain occurred after very short times in the process because of the action of HDM promoted by imposed or capillary pressures. The constant k_i depends on the mass transfer rates mainly associated with diffusional/osmotic mechanisms (Fito and Chiralt, 1997). Table 9.2 shows the kinetic parameters obtained by fitting Equations (4) and (5) to the experimental data. In PVOD treatments, both water and solute gains after very short times (k_{i0}) were positive, and solid gains were much higher than those obtained in OD treatments. This was due to a notable sample impregnation with the OS after the vacuum pulse, as expected from the sample VI capability. In OD treatments, the process advance in the water and solid gains was much lower because of the prevailing diffusional mechanisms in mass transfer rate. In fact, the k_{i0} parameters are near 0 in almost all the cases.

TABLE 9.2. Kinetic Parameters Obtained by Applying Kinetic Model for Solute and Water Gains in Orange Peel and Mandarin Peel Treated with Different Osmotic Solutions at 30°C.

Raw Material	Osmotic Solution	OD Processes			PVOD Processes			
		$k_{s0}10^2$	$k_s 10^3$	$k_w 10^3$	$k_{s0}10^1$	$k_s 10^3$	$k_{w0}10^1$	$k_w 10^3$
O	65° Brix sucrose solution	9.06	0.32	−1.67	4.30	1.53	2.01	−1.25
O	55°Brix glucose solution	—	1.59	−1.30	4.33	0.42	2.45	−1.03
O	58°Brix RGM solution	—	1.28	−2.15	4.48	−0.30	2.61	−1.56
M	64°Brix RGM solution	—	0.71	−1.98	4.59	−0.36	2.28	1.82

O: orange; M: mandarin.

$$\Delta M_w = k_{w0} + k_w\, t^{0.5} \tag{7}$$

$$\Delta M_s = k_{s0} + k_s\, t^{0.5} \tag{8}$$

Table 9.3 shows the values of the total mass loss (ΔM_t), solute gain (ΔM_s), and water loss (ΔM_w), for different osmotic treatments of orange and mandarin peels, at different times of treatment: 15 min for PVOD treatments and 3 h for OD treatments. At these times, similar solute concentration in the product liquid phase (approximately 40%) was reached in all treatments. It can be observed that in PVOD treatments, solute gains were much greater than those obtained in OD treatments, and, although a net water loss occurred in OD, water gains were observed in PVOD, in line with the effect of the vacuum pulse. The great difference in process time necessary to obtain the same sample concentration level between OD and PVOD treatments is also remarkable and reflects the notable effect of VI on the kinetics of osmotic process in these porous products. So, by applying a vacuum pulse in 15 min osmotic treatments of orange and mandarin peels, it is possible to obtain more than 40% of solid gain and 14% of water gain, thus reducing sample water activity to the same level as that reached after 3 h treatment at atmospheric pressure.

COMPOSITIONAL CHANGES IN THE PRODUCT LIQUID PHASE

Compositional changes in the fruit liquid phase in the reduced concentration (Y_w) referred to the mass fraction of water (z_w^t) [(Equation 9)] for the different treatments were analyzed by fitting a straight line to the experimental points plotted as shown in Figure 9.3. The obtained slope can be related with a pseudo-diffusion coefficient (D_e) according to Equation (9), which is an integrated Fickian equation for an infinite plane sheet geometry for short times, simplified to only one term of the series solution (Crank, 1975). To obtain the reduced concentration or process driving force ($Y = Y_w = Y_s$), equilibrium concentration (z_w^e) in each treatment was considered equal to that of the respective OS (y_w) (Chafer et al., 1998). For PVOD treatments, Y_w was referred to sample composition after the vacuum pulse ($z_w^0 = z_{w,HDM}{}^0 = 0.40$), which was calculated by considering the impregnation level (X) commented on above by applying Equations (10), (2), and (3). This was done to make the D_e values more comparable in OD and PVOD treatments. The great value of $z_{w,HDM}$, which reflects a great reduction in the process driving force promoted by the vacuum pulse, is remarkable. Assuming that no transport occurs through peel epidermis covered with natural wax, the entire sample thickness was considered in Equation (9).

$$1 - Y_w = (z^0_{W,HDM} - z_w^t)/(z^0_{w,HDM} - z_w^e) = \left(\frac{4D_e}{e^2\pi}\right)^{1/2} t^{0.5} \tag{9}$$

$$z^0_{w,HDM} = \frac{z_w^0\,(1 - x_i^t)(x_w^0 + x_s^0) + y_w x_i^t}{(1 - x_i^t)(x_w^0 + x_s^0) + x_i^t} \tag{10}$$

TABLE 9.3. Solute and Water Gains in Orange Peel and Mandarin Peel for Different Treatments at 30°C All of Which Lead to 40°Brix Products.

Raw Material	Treatment	Osmotic Solution	t (min)	ΔM_t	ΔM_w	ΔM_s
O	OD	65°Brix	180	−0.06 ± 0.02	−0.171 ± 0.002	0.129 ± 0.003
O	PVOD	Sucrose solution	15	0.58 ± 0.02	0.158 ± 0.002	0.48 ± 0.03
O	OD	55°Brix	180	−0.02 ± 0.005	−0.13 ± 0.03	0.14 ± 0.003
O	PVOD	Glucose solution	15	0.711 ± 0.008	0.183 ± 0.005	0.445 ± 0.007
O	OD	58°Brix	180	−0.052 ± 0.004	−0.19 ± 0.02	0.09 ± 0.01
O	PVOD	RGM solution	15	0.63 ± 0.002	0.200 ± 0.008	0.46 ± 0.02
M	OD	64°Brix	180	−0.152 ± 0.002	−0.209 ± 0.005	0.055 ± 0.002
M	PVOD	RGM solution	15	0.615 ± 0.007	0.141 ± 0.006	0.433 ± 0.003

O: orange; M: mandarin; RGM: rectified grape must.

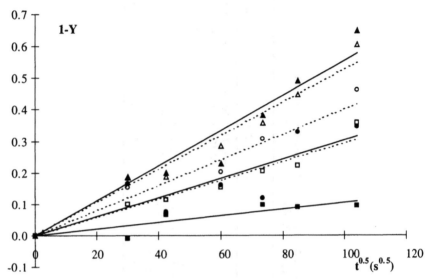

Figure 9.3 Kinetics of compositional changes in the product liquid phase in the development of reduced driving force (Y_w) in OD (open symbols) and PVOD treatments (closed symbols) at 30°C (square), 40°C (circle), and 50°C (triangle).

The obtained D_e values for OD and PVOD processes, by using different osmotic solutions, at different temperatures (30, 40, and 50°C) are compared in Figure 9.4. In general, these values were very similar in both treatments (PVOD and OD) as opposed to what was observed for other porous products, where the previous sample impregnation promoted diffusional mechanisms in the pore liquid phase (Fito and Chiralt, 1997; Fito and Chiralt, 2000). Nevertheless, this can be explained in the contribution of progressive capillary impregnation of the samples pores with the osmotic solution to the composition change in OD treatments and so, to the value of the effective diffusion coefficient (Chafer et al., 2000b).

INFLUENCE OF OSMOTIC TREATMENTS ON COLOR, MECHANICAL PROPERTIES, AND SENSORY ACCEPTABILITY

Leaving the great reduction in the process time achieved in PVOD process to one side, the notable difference in the retained liquid phase in samples processed by OD and PVOD treatments may confer different quality attributes. Quality of OD- and PVOD-treated samples (concentrated till 40°Brix)

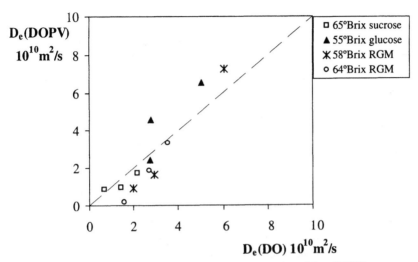

Figure 9.4 Comparison of diffusion coefficient values obtained from OD and PVOD treatments conducted on mandarin (O) and orange peel (□, ★, ▲) with different osmotic solutions and at different temperatures (30, 40, and 50°C). RGM: rectified grape must.

has been evaluated through instrumental measurements of color (from reflectance spectra), mechanical response (from a puncture test), and by sensory evaluations of different attributes. Table 9.4 shows the values of force and distance at the yield point in the puncture curve for fresh, OD- and PVOD-treated samples. It can be observed that PVOD-treated samples show the lowest force values, thus indicating a greater softening of the structure due to the treatment, which does not occur in treatments at atmospheric pressure.

Color and appearance of processed orange peel by OD are hardly affected, whereas they are sensibly affected by PVOD treatment because of the gas-liquid exchange promoted by the vacuum pulse. Gas substitution implies an increase in the homogeneity of the sample refractive index (n) and, therefore, an increase in transparency and in the sample depth for the selective light ab-

TABLE 9.4. Values of Force (F) and Distance (d) at
the Yield Point Obtained from a Puncture Test[a]
for Fresh (15°Brix) and Processed (34°Brix) Orange
Peel (Osmotic Solution: Sucrose Solution).

Samples	F	d
Fresh	7.9 ± 0.7	19 ± 5
OD treated	19 ± 3	8.9 ± 0.3
PVOD treated	15 ± 4	9.1 ± 0.6

[a]Deformation rate: 1.5 min/s; plunger diameter: 4 mm.

TABLE 9.5. Color Coordinates (L^*: clarity; C^*_{ab}: chrome; and h^*_{ab}: hue) of Fresh and Treated Samples (40°Brix) at Atmospheric Pressure (OD) and by Applying a Vacuum Pulse (PVOD) (Osmotic Solution: Sucrose, 65°Brix).

Samples	L^*		C^*_{ab}		h^*_{ab}	
	Flavedo	Albedo	Flavedo	Albedo	Flavedo	Albedo
Fresh	67 ± 2	91 ± 2	78 ± 2	31 ± 4	64 ± 3	96 ± 1
OD treated	66 ± 1	89 ± 2	76 ± 2	39 ± 7	64 ± 2	94 ± 1
PVOD treated	58 ± 3	63 ± 7	61 ± 5	33 ± 4	65 ± 3	87 ± 4

sorption. So, reflectance decreases selectively as a function of wavelength (Hutchings, 1999). Table 9.5 shows the color coordinates for fresh and treated samples of albedo and flavedo surfaces. The decrease in clarity (L^*) provoked by PVOD treatment, especially in the more porous albedo zone, where greater gas-liquid exchange occurs, can be observed, in agreement with the reflectance decrease. Chrome also decreases in PVOD-treated samples in both flavedo and albedo surfaces. No significant changes in flavedo hue are provoked by treatments, although albedo hue shifts to yellow because transparency increase allows the flavedo color to be transmitted.

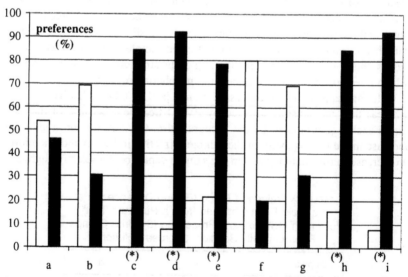

Figure 9.5 Scores reached in a preference test of samples osmotically treated with sucrose solution (65°Brix), at atmospheric pressure (white bars), and by applying a vacuum pulse (dark bars). (sample°Brix: 40%). (*) Statistically significant differences. Evaluated attributes: (a) albedo color, (b) flavedo color, (c) general appearance, (d) sample flavor, (e) sample sweetness, (f) sample bitterness, (g) sample sourness, (h) texture, and (i) overall preference.

PVOD treatments change the quality attributes of orange peel more than OD at atmospheric pressure. Product becomes softer and has different optical properties, which gives the product a juicier aspect, without losing freshness. A sensory preference test on OD- and PVOD-treated samples till 40°Brix was reached was conducted to evaluate different attributes (Figure 9.5). Significantly greater preference was obtained for PVOD samples in appearance, sample flavor, sweetness, and texture. The overall preference was significantly greater for PVOD-treated samples.

CONCLUSIONS

Orange peel processed by osmotic treatments is an interesting product for human consumption because of its natural healthy and nutritive components and the enhanced sensory properties. Pulsed vacuum treatments especially improve the product texture, sweetness, and appearance, making the product more consumer acceptable. Therefore, this kind of treatment implies a sharp reduction in process time, compared with those conducted at atmospheric pressure, because of the impregnation of the greatly porous structure of albedo with the external sugar solution. So PVOD treatments are recommendable to obtain orange peel products with better quality attributes after much shorter treatment times.

NOMENCLATURE

M_t = sample mass at time t ($t = 0, \ldots$), kg
ΔM_s = solute gain, solute kg/initial product kg
ΔM_w = water gain, water kg/initial product kg
e = sample thickness, m
Y = reduced driving force
D_e = diffusivity value, m^2/s
x_i^t = mass fraction of the component i in food at time t ($t = 0, \ldots$), i kg/total kg
x_i^e = mass fraction of the component i at equilibrium, i kg/total kg
z_i^t = mass fraction of the component i in the food liquid phase (FLP) at time t ($t = 0, \ldots$), i kg/FLP kg
z_i^e = mass fraction of the component i in the FLP at equilibrium, i kg/FLP kg
z_{wHDM}^0 = mass ratio of the impregnated solution in the initial FLP impregnated solution, kg/FLP kg
x_i^I = mass fraction of the component i in the impregnated product, i kg/total kg

x_{HDM} = mass ratio of the impregnated solution in the initial product impregnated solution, kg/total kg

y_i = mass fraction of the component i in the osmotic solution

X_1 = sample volume fraction impregnated at the end of vacuum step in VI process, m^3/m^3 initial sample

X = sample volume fraction impregnated after a VI process, m^3/m^3 initial sample

F = maximum force value in the force-distance curves, N

d = maximum distance value in the force-distance curves, mm

L^* = luminosity coordinate in the CIE-L*a*b* system

C^*_{ab} = chrome psychrometric coordinate

h^*_{ab} = hue psychrometric coordinate

Greek letters

ρ^O = bulk density of initial sample, kg/m^3

ρ^{IS} = density of the impregnating solution, kg/m^3

ϵ_e = effective porosity

ϵ_T = total porosity

γ = volume sample deformation at the end of the VI process, m^3/m^3 initial sample

γ_1 = volume sample deformation after the vacuum step in VI process, m^3/m^3 initial sample

Subscripts

OS = osmotic solution

HDM = hydrodynamic mechanism

t = values at time t

Superscripts

i = food components ($i = 1,2,...n$)

s = solutes

w = water

e = equilibrium

1 = vacuum step

2 = atmospheric step

REFERENCES

Baker, R. A. 1980. "The role of pectin in citrus quality and nutrition." In: *Citrus Nutrition and Quality,* Ed., S. Nagy and J. A. Attaway. ACS Symposium Series 143, Washington, D.C.

Barat, J. M., Chiralt, A., and Fito, P. 1998. "Equilibrium in cellular in food osmotic solution systems as related to structure." *J Food Sci.* 63:1–5.

Barat, J. M., Alvarruiz, A., Chiralt, A., and Fito, P. 1997. "A mass transfer modelling approach in osmotic dehydration." In: *Engineering and Food at ICEF.* Part I. Scheffield Academic Press Ltd., United Kingdom, pp. 77–80.

Bomben, J. L. and King, C. J. 1982. "Heat and mass transport in the freezing of the apple tissue." *J. Food Technol.* 17:615–632.

Borroto, B., Rodriguez, J. L., and Larrauri, J. A. 1995. "Chemical composition of citrus husk dietary fiber during its season." *Alimentaria.* 265:63–65.

Braddock, R. J. 1999. *Handbook of Citrus By-Products and Processing Technology.* John Wiley and Sons, Inc., New York.

Cerezal, P. and Piñera, R. M. 1996. "Carotenoides en las frutas cítricas. Generalidades, obtención a partir de desechos del procesamiento y aplicaciones." *Alimentaria.* Nov. 19–32.

Cháfer, M., Ortolá, M. D., Chiralt, A., and Fito, P. 1998. "Osmotic dehydration of orange peel." *Proceedings of the 11th Internatinal Drying Symposium (IDS'98).* Haldiki, Greece, A:886–894.

Cháfer, M., Ortolá, M. D., Martinez-Monzó, J., Navarro, E., Chiralt, A., and Fito, P. 2000a. "Vacuum impregnation and osmotic dehydration of mandarin peel." *Proceedings of the ICEF 8,* Puebla, Mexico. Technomic Publishing Co., Inc., Lancaster, PA. In press.

Cháfer, M., Gonzalez-Martínez, C., Ortolá, M. D., Chiralt, A., and Fito, P. 2000b. "Osmotic dehydration of mandarin and orange peel by using rectified grape must." In: P. J. M. Kershof, W. J Courmans, and G. D. Mooiweer (Eds): *Proceedings of the 12th International Drying Symposium.* IDS'2000. Paper No 103. The Hague. Holland. Elsevier Science B.V.

Chander, S. K., Landsdown, A. G., Lugmani, Y. A., Gomm, J. J., Coope, R. C., Gould, N., and Coombes, R. C. 1994. "Effectiveness of combined d-limonene and 4-hydroxiandrostenedione in the treatment of NMU-induced rat mammary tumours." *Br. J. Cancer* 69(5):879–882.

Chiralt, A., Fito, P., Andrés, A., Barat, J. M., Martínez-Monzó, J., and Martínez-Navarrete, N. 1999. "Vacuum impregnation: A tool in minimally processing of foods." In: F. A. R. Oliveira, and J. C. Oliveira (Eds.): *Processing of Foods: Quality Optimization and Process assessment.* Boca Raton: CRC Press, pp. 341–356.

Cohn, R. and Cohn, A. R. 1996. Subproductos del procesado de cítricos. In: D. Arthey and P. R. Ashurst (Eds): *Procesado de frutas.* Ed. Acribia, S.A. Zaragoza.

Crank, J. 1975. *The Mathematics of the Diffusion.* Lorendon Press, Oxford, UK.

Fito, P. 1994. "Modelling of vacuum osmotic dehydration of food." *J. Food Eng.* 22:313–328.

Fito, P. and Chiralt, A. 1997. "An approach to the modelling of solid-liquid operations: Application to osmotic dehydration." In: P. Fito, E. Ortega, G Barbosa (Eds): *Food Engineering 2000.* Chapman and Hall, New York, pp. 231–252.

Fito, P. and Chiralt, A. 2000. Vacuum impregnation of plant tissues. In Alzamora, S. M., Tapia, M. S., and Lopez-Malo, A. (Eds). *Minimally Processed Fruits and Vegetables.* Gaithersburg, MD: Aspen Publishers, Inc., pp. 189–204.

Fito, P. and Pastor, R. 1994. "An update on vacuum osmotic dehydration." In: G. V. Barbosa-Cánovas and J. Welti-Chaves (Eds): *Food Preservation by moisture control: Fundamentals and Applications.* Technomic Publishing Co., Inc., Lancaster, PA, pp. 351–372.

Fito, P., Andres, A., Chiralt, A., and Pardo, P. 1996. "Coupling of hydrodynamic mechanism and deformation-relaxation phenomena during vacuum treatments in solid porous food-liquid systems." *J. Food Eng.* 27:229–240.

Girard, B. and Mazza, G. 1998. "Functional grape and citrus products." Ch. 5. In: G. Mazza (Ed): *Functional Foods: Biochemical and Processing Aspects*. Technomic Publishing Co., Inc. Lancaster, PA, pp. 139–191.

Hutchings, J. B. 1999. *Food Color and Appearance*. Aspen Publishers, Inc., Gaithersburg, MD.

Larrauri, J. A., Perdomo, U., Fernandez, M., and Borroto, B. 1995. "Selection of the most suitable method to obtain dietary powdered fibre tablets." *Alimentaria*, 265:67–70.

MAPA (Ministerio de Agricultura, Pesca y Alimentación). 1991. *Market Studies of Organic Product Consumption*. 1st Quarter. DOXA S.A.

Phillips, L. R., Malspeis, L., and Supko, J. G. 1995. "Pharmacokinetics of active drug metabolites after oral administration of perillyl alcohol, an investigational antineoplastic agent, to the dog." *Drug. Metab. Dispos.* 23(7):676–680.

Ponting, J. D. 1973. "Osmotic dehydration of fruits—Recent modifications and applications." *Process Biochem.* 8:18–20.

Salvatori, D. 1998. Deshidratación osmótica de frutas: cambios composicionales y estructurales a temperaturas moderadas. Tesis doctoral, Universidad Politécnica de Valencia, España.

Salvatori, D., Andrés, A., Chiralt, A., and Fito, P. 1998. "The response of some properties of fruits to vacuum impregnation." *J. Food Process Eng.* 21:59–73.

Shi, X. Q., Chiralt, A., Fito, P., Serra, J., and Escoin, C. 1996. "Application of osmotic dehydration technology on jam processing." *Drying Technol.* 14:841–857.

Spiegel-Roy, P. and Goldschimdt, E. E. 1996. *Biology of Citrus*. Cambridge University Press.

Storey, R. and Treeby, M. T. 1994. "The morphology of epicuticular wax and albedo cells of orange fruit in relation to albedo breakdown." *J Horticultural Sci.* 69(2):329–338.

Vial, C., Gulibert, S., and Cuq, J. L. 1991. "Osmotic dehydration of kiwi fruits: Influence of process variables on the colour and ascorbic acid content." *Sciences des Aliments.* 11:63–64.

Wirthgen, B. 1994. "Consumer acceptance of organic trademarks." *Verbraucherdienst* 39(12):267–273.

Yang, A. P., Wills, C., and Yang, T. C. S. 1978. "Use of combination process of osmotic dehydration and freeze drying to produce a raisin type lowbush blueberry product." *J Food Sci.* 52:1651–1664.

PART III

SALTING PROCESSES AT ATMOSPHERIC PRESSURE

CHAPTER 10

Osmotic and Diffusional Treatments for Fish Processing and Preservation

H. BYRNE
P. NESVADBA
R. HASTINGS

HEALTH ISSUES

DIET has a strong influence on health, in particular heart disease. Current recommendations for a healthy diet include: drinking less alcohol, eating less saturated fat and cholesterol, eating less sugar and salt, eating more carbohydrates and fiber, and eating more fish (Brehm, 1993). These recommendations have been based on many years of research. The positive health benefits of fish oils were first reported in 1972 where the low incidence of heart disease in Greenland Eskimos was attributed to their high intake of marine oils (Bang et al. cited in Piggott and Tucker, 1990). Other research has confirmed these findings (Burr et al. cited in Nettleton, 1992). Cold-water fish, such as Atlantic salmon (*Salmo salar*) and herring (*Clupea harengus*), contain polyunsaturated fatty acids (PUFA), i.e., the omega-3, that help prevent heart disease. A study in the Netherlands found that people who consumed one or two fish meals a week had a lower incidence of heart disease than people who ate less fish (Brehm, 1993). Dieticians recommend two or three meals containing fish per week; however, the consumption of deep-fried, smoked, pickled, and salted fish is not encouraged (Holub, 1992). Populations who consume high levels of sodium have a higher prevalence of high blood pressure, which is a contributory factor in heart disease, stroke, and kidney damage.

ECONOMIC ISSUES

The extension of shelf life of salt-treated products is accomplished by reducing the water activity by lowering the water content and increasing the salt

111

or other solute conent of the food. The constraints to the formulation of new products are economic constraints and consumer acceptance. The products must be sellable and producible at an economical cost.

Osmotic treatments may offer these economic advantages, as was pointed out by Raoult-Wack (1994). Newly developed processing methods for dewatering and impregnation create new food products and save energy. However, their widespread industrial use has been hampered by the need to validate the microbiological status of the osmotic or salt solutions, and to resolve questions of solution management such as recycling and disposal.

CONTROLLING MICROBIAL LEVELS

The objective of the osmotic treatments is to produce products that may be stored without severe heat treatment, aseptic packaging, or freezing. Fish processors must ensure that the products they supply are safe for human consumption, not only when they leave the factory but also at the point of sale. *Listeria monocytogenes* has been isolated from many food products including soft cheeses, fish, and shellfish. The human infection route of *Listeria* spp. is probably through contaminated food although the organism is ubiquitous. In healthy people, an infective dose will cause minor flu-like symptoms; however, in the very young, old, or immunocompromised, infections can be fatal. *Listeria* spp. are a particular problem to food producers because, although they grow optimally at 30–37°C, they also multiply at approximately 0°C, and so refrigerated food containing low levels of the bacteria leaving the factory may contain an infective dose at point of use. Growth can occur in the presence or near absence of oxygen, in foods with a pH as low as 4.5, and in a high-salt environment where water activity is down to 0.9. They are, however, destroyed by high temperatures (72–75°C). It is important, therefore, that any brining solutions used to treat fish are kept clean to prevent contamination between batches. Various methods of treating brines are under investigation, including ozonation, ultraviolet light, and microwaves (Kozempel et al., 1997). It has been suggested that the most probable source of *Listeria monocytogenes* in salmon smoke houses is the raw fish (Macrae et al. cited in Embarek et al., 1997). Heinitz and Johnson (1998) found the incidence of *Listeria monocytogenes* in smoked fish and shellfish to be 14% in 1080 samples. However, the work of Embarek et al. (1997) did not confirm these findings, and they could not explain the relatively high incidence of *Listeria monocytogenes* in cold smoked salmon. Further work is required on the source of contamination in smoke houses. Failure to control contamination in food production can have serious economic consequences. A study in the United States found 5 of 37 firms producing smoked fish contaminated with *Listeria* spp. went out of business because of bad publicity, and 6 of 157 firms whose products were not contaminated also went out of business (Dillon et al. cited in Heinitz and Johnson, 1998). Temperature, pH, salt, water activity, and chemical preservatives, such as organic acids, are the most common environmental factors used to maintain and increase shelf life of food products. Low temperatures reduce bacterial growth, and salt, acetic acid, and phenolic compounds have an antibacterial action. A combination of environmental factors is usually required for effective control. Thurette et al. (1998) found that salt concentrations in the range of 2–4% were too low to limit the growth of *Listeria* spp. in cold smoked salmon; the inhibitory effect of salt depended on other factors—mainly the phenol concentration. Niedziela et al. (1998) found the combination of salt and formaldehyde to be most effective. Other methods of controlling *Listeria* spp. in food are irradiation, modified atmosphere packaging in CO_2, treatment with monoglycerides, and rinses with organic acids (Bal'a and Marshall, 1998). Water activity (a_W) is the water available for the growth of microorganisms or for chemical reactions during processing and storage in food. It is the freedom of water to move or interact with the other ingredients in the food and not just the water content that is the key to product stability (Piggott and Tucker, 1990). Water activity depends on the solvents and solutes in a system as well as the water content. If it can be reduced or inactivated, there is a point at which microbial growth is inhibited, and the shelf life of the food is increased. Salt or other products added to the dried food reduce the availability of water for deteriorative processes in food.

PRESERVATION OF FISH PRODUCT

A wide variety of fish products are on the market, including dried, salted, smoked, and pickled or marinated fish. Fish were traditionally processed to increase their storage life, although today some are processed primarily for their flavor.

DRIED FISH

It is estimated that 25% of the worlds' fish catch destined for human consumption is dried (Piggott and Tucker, 1990). Traditionally, in hot countries, fish were dried in the sun and/or wind. This was far from ideal because the drying rate could not be controlled and bacteria could multiply in the flesh. Machines are now used to give more control over the process. Factors that affect drying are atmospheric pressure, humidity, temperature, velocity, and direction of air passing over the product, and the product surface area. There are two distinct phases of drying. First, there is a constant drying rate that occurs while the product surface is wet. The energy added evaporates the water; during this phase high air flows, high air temperature, and low humidity increase drying. The second phase occurs when the water can no longer migrate to the surface as fast as the energy is absorbed. The rate of drying slows, and the product temperature rises. During this period high air temperatures increase drying by causing faster capillary diffusion within the product. Water diffuses to the surface faster in leaner and thinner fish, whereas increased salt concentration in the flesh reduces water movement. The final water content of the product depends on the relative humidity of the air in the dryer. The lowest water content can be achieved by using a vacuum dryer (Piggott and Tucker, 1990).

SALTED FISH

Curing fish by using salt is an old preservation technique. It can be achieved by brining or dry salting. Salt denatures muscle, causing proteins to become insoluble and lose their associated water. To dry salt fish, the fillets are covered with salt, and the brine liquor produced is allowed to escape. This technique is most suitable for low-fat fish such as cod (*Gadus morhua*). Traditionally, fatty fish such as mackerel (*Scomber scombrus*) are brined to ensure a more even salt distribution. Brining is immersing fish in brine before smoking. A common working strength for batch brining is 80° (100° corresponds to a saturated solution containing 264 g salt per 1 kg water at 16°C). Dry salted products can be hard or light cured. A hard cure has a water activity of 0.75–0.85 and a light cure of 0.85–0.9 (Piggott and Tucker, 1990). The shelf life of the light cure is short (a few days), especially at high temperatures, whereas hard cured fish will last a few weeks under similar conditions. Salt may act as a prooxidant in fish flesh with a subsequent increase in lipid oxidation and reduction in vitamin content (thiamine levels are decreased 14–24% in salted fish products (Daun cited in Piggott and Tucker, 1990). Several factors affect the absorption of salt into the fish; these include the grain size and purity of the salt, the freshness and fat content of the fish, the thickness of the flesh and the skin, and the temperature.

SMOKED FISH

Smoked fish can be either hot smoked, where they are cooked and smoked, or cold smoked, which are simply "cured."

Hot smoked fish include, for example, rainbow trout (*Oncorhynchus mykiss*) (Mills, 1978). In the production of hot smoked trout, whole gutted fish post rigor mortis are used as salt uptake is reduced in prerigor muscle. Fish of a similar size are selected for uniform salt uptake in the batch; a final concentration of 3% salt is required. The uptake of salt depends more on the curing time than the temperature. A brine strength of 80° (28 g/100 g) is often used. To ensure adequate salt uptake, the fish should be completely immersed and there must be three parts brine to one part fish; higher ratios lead to brine depletion. Once the salt in the fish and brine has reached equilibrium, the fish are drained and smoked. The smoking process warms the fish slowly to 30°C to allow the skins to dry. The temperature is then raised quickly to 50°C for 0.5 h and then to 80°C for 0.75–1 h or until they are cooked, which depends on factors such as atmospheric temperature and humidity. The end product must be cooled before packing at <4°C. Approximate storage times are 1 week at 4°C, 2–3 days at 5–10°C, and 12 months frozen at −30°C.

Cold smoked fish include, for example, salmon (Bannerman and Horne, 1980). The traditional method of producing cold smoked salmon is to dry salt the fillets. Brining has been used, but this extends the drying time in the kiln; however, some fish processors use injection-brining techniques. Chilled or thawed frozen salmon can be used; it is preferable to use high-fat content fish that were gutted soon after slaughter because this improves product quality. To increase salt uptake, cuts that just penetrate the skin are made or, alternatively, small circles of skin removed. The thinner tail portions of the fillets are lightly salted to achieve uniform salt distribution. The fillets are then left on salt for up to 36 h, depending on size of the fish, the initial quality, and the fat content. A 2.5% salt content in the finished product is required (i.e., 12 h for 4-kg fish). Salted fillets are then washed to remove excess salt. It is recommended that the fillets are then immersed in 30° brine for 0.5 h to even out the salt distribution. The fillets can have lost up to 9% of their weight during this process. The fillets are then drained and kept at 2°C for at least 4 h to reduce pellicle formation. They are then placed in the smoker. Hardwood sawdust is used for smoking; initially, the temperature is at 27°C and the fillets are smoked gently to prevent overdrying. This can be for 4–10 h, depending on the size and fat content of the fillets. The kiln temperature is then raised to 33°C for the last 15–20 min to bring the oil to the surface and give the fillets a shiny appearance. Fillets from 4- to 5-kg fish of high-fat content need approximately 6 h. After approximately 5 h, the fillets are usually sufficiently smoky, so the chimney is shut and the heater and fans are left on to continue the drying process. The fillets should lose another 7–9% of their weight in the kiln. Once cool, the sides can be sliced and vacuum packed. Kept at <4°C, the shelf life is 5–10 days; if frozen at −30°C, they can be kept for 6 months.

PICKLED/MARINATED FISH

Marinated herring (McLay, 1982) are a traditional food consumed in several North European countries. The manufacturing process was developed to increase the storage lives of fish and is used mainly for herring and shellfish. It involves preserving in an acidic solution (usually acetic acid; sometimes citric or tartaric acid) and salt. The action of bacteria and enzymes are retarded.

The solution can be flavored with sugar and spices. The process alters the muscle proteins, and the flavor and textural properties that develop have resulted in the preserved fish becoming a product in its own right (Rodger et al., 1984). Salt and acid reduce the water content in the fish and the required texture is achieved by altering the final levels of salt and acid in the tissue. The fish are first washed, and if necessary, put in 10% brine for 1 h to "firm up." They are then gutted, filleted, and rinsed in 5% brine to remove any blood. The prepared fish are then immersed in a "strong" solution of acetic acid and salt in vats in a cool room for up to 3 weeks. The ratio of fish to liquid is important. The higher the ratio of fish to solution the more concentrated it has to be. Traditional methods in open vats require 1.0–1.5:1 fish to liquid. A higher ratio would require a solution so concentrated that the fish would float. To achieve the required final concentration of acetic acid in the fish (2.5%), the initial concentration of the solution should be 4% acetic acid and 10% salt. In closed vats in which the fish cannot float, the ratio of fish to liquid can be increased to 2.3:1. The initial solution should contain 7% acetic acid and 14% salt. The herring and the solution are put into the vats a little at a time to prevent rapid dilution. To ensure complete mixing and to stop the fish sticking together, open vats are stirred and closed vats are rotated. Vats are regularly inspected and lost liquid is replenished. The fish absorb the acetic acid and salt from the liquid until equilibrium is reached, approximately 1–3 weeks depending on room temperature. They can be left in these vats for up to 6 months at 3°C if required. After marinating, the flesh is white, opaque, and firm. Discolored fish are discarded, and those remaining are packed in jars with pickle or sauces. The ratio of fish to covering liquid should be between 1:1 and 2:1, and the liquid should contain 1–2% acetic acid and 2–4% salt. To reduce the acidic taste, it is possible to substitute citric acid for some or all of the acetic acid in the bottling process. The pH should not exceed 4.5 to achieve storage life of several months at 4°C (McLay, 1982). By heat pasteurizing or sterilizing the products, the shelf life is extended much further. The salt content of the final marinade should be approximately 2% and acid content approximately 1.5%.

Some speciality semipreserves akin to marinades are Scandinavian anchovies (made of sprat or herring, not anchovies, preserved in spiced brine). A similar German product "Anchosen" and spiced herring called "tidbits" or "gaffelbidder" contain sugar and salt. There are also herring products preserved by fermentation and salt.

FREEZING IN OSMOTIC SOLUTIONS

The primary reason for freezing in osmotic solutions is the very high rate of heat removal. Other benefits are that individual products do not stick together as in air blast freezing. Diffusion of solutes into the products can be minimized by forming a crust of ice on the product surface or by adding sugar to the salt solution. Conversely, allowing diffusion of solutes into the product opens possibilities for quality and shelf life improvement and product development (Lucas et al., 1997).

MEASURING LEVELS OF SALT

The salt concentration in the flesh of smoked fish is routinely monitored because it has implications for microbiological safety. A sample is selected from the fish where the concentration of salt is lowest. The salt content is then determined in g salt/100 g of flesh. The following three methods are recommended by the Association of Official Analytical Chemists for determining salt in fish (AOAC, 1995).

VOLUMETRIC METHOD

This method relies on a chemical reaction of chloride ions from the salt in the food with silver ions in the added reagent. For this to occur the food has to be acidified. The reaction forms insoluble silver chloride, which can then be detected chemically by titration with ammonium thiocyanate, and a chemical indicator, which turns the solution light brown.

POTENTIOMETRIC METHOD

This method measures the potential difference between two electrodes immersed in the sample solution that is inversely proportional to its salt content. Water and acid are added to the fish homogenate so that the chloride ions in the food react with the added reagent silver nitrate. The voltage across the solution is measured as the reagents are added in small aliquots. A graph can then be constructed to show the voltage change with the addition of the reagents. The salt content can then be deduced by using a previously constructed calibration curve.

INDICATING STRIP METHOD

The previous methods require laboratory premises, time, and expertise. A simpler, although less accurate, method of measuring salt that has been available for use in the food processing industry for many years is the indicating strip method. The QUANTAB Chloride Titrator is a thin chemically inert plastic strip. It is laminated with absorbent paper capillary columns impregnated with brown silver dichromate. When the strip is placed in an aqueous solution of the food to be tested, the fluid rises in the columns by capillary action. Chloride in the solution reacts with the silver impregnated in the column to produce a color change to white, insoluble silver chloride. When the capillary column is completely saturated, the top of the column turns dark blue to indicate completion of the test. The length of the white column is proportional to chloride concentration in the test solution. A numbered scale is read at the top of the white color change and converted to percent salt by use of calibration table in pack (Vander Werf and Free, 1970).

MODELING LEVELS OF SALT

To determine salt uptake and distribution in fish muscle as a function of processing time, it is necessary to know the salt diffusivity. Crank defines diffusion as "the process by which matter is transported from one part of a system to another as a result of random molecular motion" (Peters, 1971). Early attempts to model salt diffusion in fish (Del Valle and Nickerson, 1967) reported that the diffusivity depended on the salt concentration in the muscle and temperature. They used a constant salt diffusion coefficient to describe salt diffusion in swordfish (*Xiphaias gladius*) muscle. Peters (1971) used Ficks law to model salt diffusion in cod muscle, and Nesvadba (1990) used the same law to model the diffusion of brine injected into fish fillets by an array of hypodermic needles, a proess used by companies for fish and ham. Rodger et al. (1984) investigated the diffusion properties of salt and acetic acid in herring. They found that total uptake of both acetic acid and salt depended more on curing time than temperature. The acetic acid diffused more rapidly than the salt into the herring, and the factors that had an effect on the diffusivity were the initial water and fat content, variation in the permeability of the skin, and biological variation. The effect of fat on the diffusivity was as expected—the higher the fat content in muscle the greater the resistance offered for an aqueous solute to transfer (Schwartzberg and Chao, 1982). Therefore, the type of species will have an effect on the uptake of salt in fish. Fattier fish, such as mackerel, will have a lower diffusivity than lean fish, such as cod. Skin structure also has an effect because the skin inhibits salt diffusion as does the layer of flesh immediately beneath the skin, which has a high fat content. Spawning condition has a large effect on the fat content of fish and so will have a significant effect on salt diffusivity. It is possible to determine the fat levels in a fish by using a Torry Fat meter. Numerical methods can be used to simulate the diffusion processes where the diffusion coefficient is concentration dependent (Crank cited in Wang et al., 1998). Wang et al. (1998) produced a finite difference model to predict the salt diffusivity in Atlantic salmon and to evaluate the salt uptake in post rigor muscle at 10°C, 20% w/v. They validated the model with experimental data. Salt diffusivity was found to be salt concentration dependent. This could be explained by the gradual denaturation of the protein during salting. As the salt concentration increased, the cell structure gradually degraded, resulting in less resistance to diffusion at the higher salt concentrations (Wang et al., 1998). This leads to the conclusion that there will be a difference in the diffusivity in fresh and previously frozen fish because ice formation and melting during the freeze and thaw cycle damages the cell structure, reducing resistance to salt diffusion into the cell and resulting in higher salt diffusivities. It is possible to detect whether fish has been previously frozen by using a Torrymeter Fish Freshness meter.

DEVELOPMENT OF NEW PRODUCTS USING OSMOTIC TREATMENTS

The most important reason for having osmotic treatments is their industrial exploitation. Therefore, it is important to maintain liaison with the food-processing industry. However, from the approaches made to the fish processors in the Northeast of Scotland, there is apparent certain skepticism that the osmotically treated products would be suitable for the European market. This scepticism may be overcome by:

- showing ways of using parts of fish that would otherwise be used for fish meal, which means upgrading fish waste or underutilized species. This may involve comminuting the fish and reconstituting it as thin slices.
- developing protoype products that could be shown to the food industry
- formulating snack foods in which the salt and sugar levels are comparable with those in accepted successful snack foods such as potato crisps (0.5–2.5% salt, 5% sugar, 0.3% vinegar), salted peanuts (1% salt), or anchovies (6% salt)

VARIATION AND MEASUREMENT OF TEXTURE

Osmotic treatments can regulate both the main sensory attributes: food texture and flavor. In the raw state, a considerable array of flavors and textures is available among fish species as caught and processed (Whittle, 1990). These range from the more resilient, tough, and rubbery textures of squid, octopus, and abalone, through the firm textures of tuna, swordfish, shark, and halibut, or to the fatty species of herring and pilchard. Some of the deep water species, such as smoothead, are very soft and sloppy. Salting and marinating add a further dimension to the texture and flavor of the products.

With the decrease of water content of the product, there is a tendency for the texture to become tougher. This could be overcome by selecting softer fish species as the starting material. Osmotic treatments thus have considerable possibilities to modify the naturally occurring textures and flavors and to introduce new ones, by incorporating flavoring compounds in the osmotic solutions.

During development of new products, it is necessary to monitor their texture. Much work has been done on correlating sensory and instrumental methods of texture characterization (e.g., Rao and Gault, 1990). However, the inhomogeneous nature of fish products is an obstacle to rheological measurements. The sensory test is still the most reliable means of measuring the texture as perceived by the consumer (mouthfeel). Methods that measure the human response by instrumental means may be the way forward (McStay and Yang, 1995).

SOURCES OF FISH FOR OSMOTIC TREATMENTS

With the North Sea being "overfished," the catches are becoming very mixed (up to 15 species may be present in one haul) and increasingly include as bycatch some underutilized and deep-water species. These are usually discarded or destined for fish meal. Osmotic treatments could provide a route by which these species could be used for human consumption.

Another possibility is using fish flesh that remains on frames after filleting. Such flesh is recovered mechanically and is, therefore, in comminuted form. It may be possible to reconstitute the mince in the form of thin slices, similar to some types of potato-based snack foods (McLay, 1970).

HYGIENIC CONSIDERATIONS IN OSMOTIC TREATMENTS

Continuous systems are probably not feasible from the hygienic point of view because of the need for very large holding tanks to achieve the required throughput. For example, if 100 kg of fish are to be treated per hour, and the osmotic treatment takes 15 h, then a holding tank of more than 4500 kg is required (for the ratio 3:1 solution to fish).

A batch system (Yang and LeMaguer, 1992) would comprise a holding tank covered with a lid to prevent airborne contaminaton. A dosing system would maintain the desired concentraton of the osmotic solution (in conjunction with an on-line monitoring sensor for solute content). The soluton would be circulated through a UV and/or ozone sterilizer, followed by a coarse filter to remove food debris. After that, a fine filter could be introduced to remove small particles. Fine filters could also be used to remove bacteria, but not viruses. The pore size would have to be 0.2 μm because some bacteria (microbacteria) are not captured by a 0.5 μm pore size filter. All filters would have to be cleaned or renewed periodically. Tangential filtration systems could also be investigated because they are less likely to block.

Apart from the difficulties in solution management, osmotic treatments have the potential to be very hygienic. This is due to the simplicity of the installation and surfaces that are easy to clean. Any industrial application of the osmotic treatments would have to be subjected to a full HACCP.

CONCLUSIONS

It is important for fish processors to be aware that salt levels in processed fish should be sufficient to protect it microbiologically but not so high that it causes health problems. The beneficial PUFAs are the lipids most involved in the development of rancidity, and salt in fish may also act as a prooxidant. Processors need also to be aware that salt diffusion varies with the condition of the raw material, pre/post rigor or previously frozen fish. The fat content of fish varies with species and within species because of spawning condition and so each batch of fish needs to be assessed before processing. In fish and fish products at the moment the osmotic and diffusional processes are part of operations in which only slight dehydration takes place. There appears, therefore, to be some scope for extending the dehydration to lower water contents. This must be done while having in view the objective of developing a marketable food product. Once this is in sight, then all the techniques of mathematical modeling, characterization of products, and microbiological testing that are available can readily be applied to improving and scaling up the osmotic processes.

REFERENCES

AOAC. 1995. *Official Methods of Analysis.* 16th Edition. Association of Official Analytical Chemists, Arlington, VA.

Bal'a, M. F. A. and Marshall, D. L. 1998. Organic acid dipping of catfish fillets: Effect on colour microbial load and *Listeria monocytogenes. Journal of Food Protection,* 61(11):1470–1474.

Bannerman, A. and Horne, J. 1980. Recommendations for the Preparation of Smoked Salmon. Ministry of Agriculture Fisheries and Food, Torry Research Station. Torry advisory note No. 5, HMSO, Edinburgh.

Brehm, B. A. 1993. *Essays in Wellness.* Harper Collins, New York.

Del Valle, F. R. and Nickerson, J. T. R. 1967. Studies on salting and drying in fish. I Equilibrium considerations in salting. *Journal of Food Science,* 32:173–179.

Embarek, P. K. B., Hansen, L. T., Enger, O., and Huss, H. H. 1997. Occurrence of *Listeria* spp. in farmed salmon and during subsequent slaughter: Comparison of Listertest™ Lift and the USDA method. *Food Microbiology,* 14:39–46.

Heinitz, M. L. and Johnson, J. M. 1998. The incidence of *Listeria* spp., *Salmonella* spp., and *Clostridium botulinum* in smoked fish and shellfish. *Journal of Food Protection,* 61(3):318–323.

Holub, B. J. 1992. Potential health benefits of the omega-3 fatty acids in fish. In: *Seafood Science and Technology,* ed. Bligh, E. G. Fishing News Books, Oxford.

Kozempel, M., Scullen, O. J., Cook, R., and Whiting, R. 1997. Preliminary Investigation using a batch flow process to determine bacteria destruction by microwave energy at low temperature. *Lebensm.-Wiss. u.-Technol.,* 30:691–696.

Lucas, T., Sereno, A., Billiard, F., and Raoult-Wack, A. L. 1997. Immersion of foods in concentrated aqueous solutions at low temperatures: A process for chilling, freezing and/or formulating. *Engineering & Food at ICEF7,* ed. Jowitt, R. Academic Press Ltd., Sheffield, Part 2, F5–F8.

McLay, R. 1970. The development of new fish products. *Proc. Conference on Fish Marketing,* London, 26–27. Torry Memoir No. 357.

McLay, R. 1982. Marinades. Ministry of Agriculture Fisheries and Food, Torry Research Station. Torry advisory note No. 56, HMSO, Edinburgh.

McStay, D. and Yang, L. 1995. A video based system for food texture analysis. In: *Sensors and Their Applications,* VII, ed. Augousti, A. T. *Proceedings of the Seventh Conference on Sensors and Their Applications,* Dublin, Ireland, 10–13 September, 1995. Institute of Physics Publishing, Bristol, 349–353.

Mills, A. 1978. Handling and Processing Rainbow Trout. Ministry of Agriculture Fisheries and Food, Torry Research Station. Torry advisory note No. 74, HMSO, Edinburgh.

Nesvadba, P. 1990. Diffusion of Salt Injected into Fish. Torry document No. 2316.

Nettleton, J. A. 1992. Seafood nutrition in the 1990's: Issues for the consumer. In: *Seafood Science and Technology,* ed. Bligh, E. G. Fishing News Books, Oxford.

Niedziela, J. C., MacRae, M., Ogden, I. D., and Nesvadba, P. 1998. Control of *Listeria monocytogenes* in salmon; Antimicrobial effect of salting, smoking and specific smoke compounds. *Lebensm.-Wiss. u.-Technol.,* 31:155–161.

Peters, G. R. 1971. Diffusion in a medium containing a solvent and solutes with particular reference to fish muscle. Ph.D. Thesis, University of Aberdeen.

Piggott, G. M. and Tucker, B. W. 1990. *Seafood: Effects of Technology on Nutrition.* Marcel Dekker Inc., New York.

Rodger, G., Hastings, R., Cryne, C., and Bailey, J. 1984. Diffusion properties of salt and acetic acid into herring and their subsequent effect on the muscle tissue. *Journal of Food Science,* 49(3):714–720.

Rao, M. V. and Gault, N. F. S. 1990. Acetic acid marinading—The rheological characteristics of some raw and cooked beef muscles which contribute to changes in meat tenderness. *Journal of Texture Studies,* 21(4):455–477.

Raoult-Wack, A.-L. 1994. Recent advances in the osmotic dehydration of foods. *Trends in Food Science and Technology,* 5(8):255–260.

Schwartzberg, H. G. and Chao, R. Y. 1982. Solute diffusivities in leaching processes. *Food Technology,* 36(2):73–86.

Thurette, J., Membre, J. M., Ching, L. H., Tailliez, R., and Catteau. 1998. Behaviour of *Listeria* spp. in smoked fish products affected by liquid smoke, NaCl concentration, and temperature. *Journal of Food Protection,* 61(11):1475–1479.

Vander Werf, L. and Free, A. H. 1970. Rapid and convenient salt measurement in meat, fish and cheese. *Journal of the AOAC,* 53(1):47–48.

Wang, D., Correia, L. R., and Tang, J. 1998. Modelling of salt diffusion in Atlantic salmon muscle. *Canadian Agricultural Engineering,* 40(1):29–34.

Whittle, K. J. (1990). Restructured fish products. Torry document No. 2407.

Yang, D. C. and LeMaguer, M. 1992. Osmotic dehydration of strawberries in a batch recirculation system. *Journal of Food Quality,* 15(6):387–397.

CHAPTER 11

Problems Related to Fermentation Brines in the Table Olive Sector

A. GARRIDO-FERNÁNDEZ
M. BRENES-BALBUENA
P. GARCÍA-GARCÍA
C. ROMERO-BARRANCO

INTRODUCTION

Most commercial preparations of table olives require a salt solution, brine, for storage or for fermentation (Fernández Díez et al., 1985). Brines have different characteristics, depending on the type of olives and on processing conditions.

There are three main types of table olive (Garrido et al., 1995): green olives (mainly Spanish or Seville olives), natural black olives, and those that have been darkened by alkaline oxidation and are known as ripe or California olives.

Green or Spanish-style olives are processed as follows. First, they are treated with a NaOH solution known as lye, which penetrates two thirds of the way into the flesh. They are then treated with freshwater to remove any excess alkali, and, finally, when the last washing water is removed, the brine is added. Different types of brine are used, but they normally have a NaCl (salt) concentration ranging from 5 to 6% (w/v) during the fermentation process (Fernández Díez et al., 1985). After this stage, the salt content is raised to 8% to ensure the stability of the product throughout the rest of the storage period. When the olives are packed, the brine is discarded, and fresh brine that is prepared from food grade salt and lactic acid is added.

Ripe natural black olives are brined directly after washing, and the NaCl content in the brine is higher than for green olives, traditionally being approximately 8 to 10% (w/v), although it can be even higher in the case of some specific brining processes (Fernández Díez et al., 1985). As for green olives, these brines are discarded on packing and replaced by fresh ones prepared with food grade salt.

The traditional method used for the preservation of ripe olives is storage in brine. Because fermentation of the olives is not important, the main aim of this process is simply to preserve them before darkening (Brenes Balbuena et al., 1986). The storage period lasts from 2 to 12 months, after which the storage solutions are discarded. The olives are then darkened by a process consisting of several lye treatments followed by water washes during which air is bubbled through the solution of olives suspended in brine. The salt content of the brines finally used for packing the olives is markedly lower than that of the brine used for storage and ranges from 2 to 5%.

Other processes that take place in the preparation of olives of any of the above types, such as pitting, stuffing, etc, produce certain amounts of saline solutions that are also poured away. The resulting pollution can be considerable, especially in the case of green (Spanish) olives, which are the type with the most different forms of commercial presentation. Second to these for pollution come ripe olives, but pollution is negligible in the case of natural black olives, because they are commercialized only as whole or plain olives and never stuffed or pitted. The salt content and likelihood of pollution of these liquids depend on the degree of mechanization of the processes used and the type of factory, among other factors.

CHARACTERISTICS OF DIFFERENT TYPES OF TABLE OLIVE PROCESSING WASTEWATERS

The most-polluting wastewaters obtained in the processing of green olives are produced in the so-called "aderezo," aimed at removing the natural bitterness of the fresh fruits. Research conducted to reduce the pollution caused by this residue has developed technology that permits reused lye and reduces the number of water washes to only one or two (Castro Gómez-Millan et al., 1981). However, the treatment of these wastewaters is very difficult (García García and Garrido Fernández, 1984), and they are simply poured into evaporating ponds. Residues of NaOH and organic matter accumulate in these ponds and may be a potential hazard in the future. The only wastewaters not properly controlled in the green, Spanish-style processing are the fermentation brines, which are produced mainly at the packing factories. Because these are close or even within villages, they have no possibility of eliminating their liquid residues by evaporation in ponds.

Values of the main physicochemical characteristics of such fermentation brines important from the processing point of view are given in Table 11.1. This table includes other parameters that can illustrate polluting load (Brenes Balbuena et al., 1988a; 1988b; Garrido Fernández et al., 1992). In general, the discarded brines are characterized by their moderate lactic acid contents, low pH, and high mineral and organic dissolved solids. Their chemical oxygen demand (COD) and biological oxygen demand at 5 days (BOD_5) have values of approximately 30 and 25 g O_2/L, respectively.

TABLE 11.1. Physicochemical and Polluting
Characteristics of Green, Spanish-Style Table Olives.

Characteristics	Approximate Value
PH	3.9
NaCl (g/L)	90.0
Titratable acidity (g lactic/L)	7.0
Polyphenols (g tannic acid/L)	6.0
COD (g O_2/L)	30.0
Total solids in solution (g/L)	120.0
Volatile	2.0
Fix (mineral)	100

The storage brines of ripe olives have similar characteristics. Consequently, the problems they cause are of the same type. Wastewaters produced during the oxidation step are lyes and washing waters, which are eliminated by evaporation in shallow ponds. Sometimes, waste brines are also poured into these facilities. A great effort has been devoted to finding alternative storage methods or purification procedures. However, none of them has been able to prevent pollution from these solutions.

Brines from naturally black ripe olives have a polluting charge higher than saline solutions of green olives. Until now, no attention has been paid to controlling the waste brines, in particular because of their unimportance.

In Spain, approximately 70% of table olives are prepared as green, Spanish-style, 30% as ripe olives (by oxidation), and only a small percentage as naturally black olives (Garrido Fernández, 1994). Therefore, most problems in the discharge of brines are for the green, Spanish-style olive-packing factories. Because treatments to regenerate green, naturally black ripe, and ripe olives can be similar, this overview will deal mainly with the regeneration and reuse of green, Spanish-style olive fermentation brines.

STRATEGIES TO PREVENT POLLUTION CAUSED BY FERMENTATION OR STORAGE BRINES

Strategies to prevent pollution from brines can be diverse, depending on the type of table olive (Garrido Fernández, 1975) because, as commented on above, the processing conditions differ significantly from one type to another. Strategies will be described for the different types of increasing product importance.

As already explained, fermentation brines from naturally black ripe olives are the heaviest polluted because of their high NaCl and organic loads (Garrido Fernández, 1975). They differ from ripe and green olive brines in the kind of polyphenols in solution. In this case, the main polyphenols are anthocyanins, which color brines are purple to pink, depending on the pH of the solution.

The color is very intense and easily detectable although produced in a low proportion. They are produced in a rate of only 0.5 L/kg of fruits.

The brine used in the final packing of naturally black ripe olives is not required to have special conditions of color, transparency, etc. So, if the fruit is sweet enough, the fermentation brine can be reused in the final packing almost without treatment. A simple filtration to eliminate the solids in suspension will be sufficient.

The product needs pasteurization or addition of preservatives to guarantee stability during its shelf life. This treatment is also required when packing is with fresh brine. Then, brine reuse will not introduce any additional distortion in the naturally black ripe olive elaboration process.

Storage brines of ripe olives have similar characteristics to those of green olives. In California, salt discharges into evaporation ponds caused, after several years of storage, contamination of aquifers. The industry, therefore, had to find and use alternative, acidic, solutions without salt, for storage. Most of the table olive factories in this state are now using solutions of lactic acid, acetic acid, or a mixture of both to maintain the olives before oxidation.

This technological change solves the problem only partially, because the acidic solutions have a higher organic loads than the former brines. Nevertheless, it is appropriate for the specific environmental problems in California and fulfill the requirements of the administration in that state. In Spain, this type of olive is usually stored in brine, which is used as neutralizing solution after the alkaline treatment in the darkening step (Brenes Balbuena et al., 1986).

Because ripe olive (by oxidation) production is still relatively low in Spain, no great problems are posed currently by this type, although they will increase as the production rises.

The high volume and polluting loads of fermentation brines of green, Spanish-style table olives are of great concern in Spain. As can be deduced from Table 11.1, the brine characteristics are far from those required for final packing. Precisely, for this reason such solutions have traditionally been discarded.

Such action was favored by the absence of pollution control, but the situation has been slowly changing over the last 20 years. Pollution control authorities are progressively encouraging the industry to adopt measures to reduce the impact of wastewaters. The sector has responded by introducing some technological changes.

REUSE OF THE FERMENTATION BRINES OF GREEN, SPANISH-STYLE TABLE OLIVES

The green table olive-packing industry produces large amounts of wastewaters that can range from 2 to 15 L/kg. Physicochemical treatments of the

TABLE 11.2. Sources of Pollution in the Green Table Olive-Packing Industry and Their Relative Polluting Load and Volumes (Garrido Fernández et al., 1992. Reproduced with permission from *Grasas y Aceites*).

Source of Pollution	Contribution (%)	
	Polluting Load (COD)	Volume
Fermentation brines	70	22
Fruit washing	17	45
Glass containers washing	0.2	11
Others	12.3	22

combined wastewaters of these factories resulted in discouragingly low removal of organic load (García-García et al., 1990).

A study related with the characterization of pollution sources yielded the data showed in Table 11.2.

As can be observed, the fermentation brines contribute a small part of the total volume of wastewaters produced in a packing factory, whereas their load is outstanding. Their segregation could significantly abate overall pollution. The best way of removing them from the pollution problem is by promoting their regeneration and reuse in packing.

A packing brine should be colorless, perfectly transparent, and without any strange odor. In addition, salt and acid content in packing brines should be lower than in the previous storage solutions. Thus, treatments to regenerate fermentation brines must achieve the aims listed in Table 11.3.

Separation of suspended solids and fat is not a problem. There are numerous systems in the market that work properly with these brines. Thus, the research has been focused mainly on procedures able to remove color and polyphenols while retaining lactic acid and salt. Of the different possible systems, special attention was paid to active charcoal adsorption and to ultrafiltration.

TABLE 11.3. Aims that Must Be Achieved by Any Regeneration Treatment Applied to Green, Spanish-Style Olive Brines.

1. Totally eliminate solids in suspension.
2. Remove the fermentation brine color as completely as possible.
3. Decrease the concentration of polyphenols.
4. Decrease, if possible, the combined acidity.
5. Maintain, or at least minimize, lactic acid losses.
6. Maintain the NaCl content.
7. Maintain the pH as close as possible to original values.

Advantages of reuse are obvious. It can, potentially, prevent approximately 70% of pollution and at the same time save costs by using the lactic acid produced during the fermentation and the salt that already exists in the solution. In fact, changing the brine does not make any sense if it must be replaced by another solution that needs the addition of the same chemicals.

REGENERATION BY ACTIVE CHARCOAL ADSORPTION

The main troubles found during the regeneration of brines with active charcoal were related to the relatively unspecific action of the adsorbent. When using this procedure, color and lactic acid are adsorbed at the same time, with the consequent increase of pH. Trials are needed to find a type of adsorbent that removes the color selectively, eliminating as little acid as possible.

The various active charcoals on the market have very different behavior. In experience, it is difficult to deduce their behavior simply by reading their characteristics. Selection of the best active charcoal requires previous experiments with each type or (preferably) the obtaining of their isotherm adsorption curves.

Usually, only a few of them fulfill all the required conditions. We have found that amounts of approximately 5 g/L of these can be enough to produce sufficient decoloration for reuse of the fermentation brines. The effects of several active charcoals on some characteristics of green, Spanish-style olive brines can be observed in Table 11.4.

As observed in Table 11.4, the GA type has adsorption characteristics good enough to produce a practically colorless solution while adequately maintaining the pH values and acidity content. Recently, other active charcoals have

TABLE 11.4. Effects of Diverse Active Charcoals (5 g/L) on the Main Physicochemical Characteristics of the Green, Spanish-Style Table Olive Brines (Brenes Balbuena and Garrido Fernández 1998. Reproduced with permission from *Grasas y Aceites*).

	Characteristics			
	Color $(A_{440}-A_{700})$	Polyphenols (g Tannic Acid/L)	Titratable Acidity (g Lactic Acid/L)	pH
Initial brine	0.64	3.54	7.50	4.20
Treated brines				
GA[a]	0.15	2.06	7.00	4.20
EA[a]	0.57	2.51	7.00	4.25
F-400[a]	0.44	2.13	6.70	4.22
GPE[a]	0.48	2.06	6.90	4.23

[a]Active charcoal types.

been introduced to the market. Some of them may have even better adsorption conditions.

Separation of adsorbent can be achieved easily by conventional filtration or (better) cross-filtration, using ceramic membranes. This material can be cleaned with alkaline or acidic solutions as well as being heat sterilized.

The procedure has the advantage of permitting active charcoal dose rate depending on the final use planned for the regenerated brines. Costs can then be fairly well controlled.

REGENERATION BY ULTRAFILTRATION

The critical point in the application of this procedure is the cutoff membrane pore size. Brine color is removed with relatively small pore size. The best results were obtained by using membranes of 4000 daltons or smaller. The chemical composition of membranes is also vital. Polyphenols from brines can react with membrane material, causing rapid flux decrease. Fouling is also easy. In both cases, there is a dramatic diminution of the flux.

Normal use also requires frequent cleaning, which takes about 1 h. This operation is determinant. It must be conducted very carefully because, otherwise, there may be a marked reduction in filtration capacity. Normally, after the cleaning operation, the flux is not recovered 100%. The time that the membranes can be used has not been tested experimentally. Makers assure that they might last at least 7 years.

Flux of permeate depends on working pressure, salt content, organic matter concentration, temperature, and pretreatment. In some cases, the last may be determinant because some pretreatments can double the flux. With respect to temperature, the upper limit is approximately 40–45°C, because higher temperatures increase permeate browning.

The effect of ultrafiltration on the main physicochemical characteristics of brines can be observed in Table 11.5. In short, the most appreciable effects

TABLE 11.5. Effect of Ultrafiltration (2000 daltons) on the Physicochemical Characteristics of Green, Spanish-Style Olive Brines (Brenes Balbuena et al., 1988a. Adapted with permission from *J. Food Sci.*).

Characteristic	Initial Brine	Regenerated Brine
pH	3.70	3.68
Titratable acidity (g lactic/L)	8.10	7.50
Combined acidity (mN)	63.00	54.00
Salt content (g/L)	80.00	78.00
Polyphenols (g tannic acid/L)	2.10	1.45
Color (A_{440}–A_{700})	0.44	0.06

are a drastic diminution in color, a slight reduction of titratable acidity, a very slight reduction of combined acidity, and a practically equal salt content with respect to the original.

RESULTS OBTAINED BY REUSING REGENERATED BRINES IN PACKING

As previously mentioned, concentrations of salt and titratable acidity in final packing brines are lower than those required in fermentation. Regenerated brines must, therefore, be diluted when reused. Their percentages range from 30 to 70%, depending on the client's requirements.

The only drawback found in reuse is that pH is slightly above the traditional one (approximately 0.2 units above). This increase is enough, however, to cause instability during the product shelf life (Brenes Balbuena et al., 1989 and 1990). Thus, use of pasteurization is necessary to stabilize the final product (Table 11.6). A few years ago, this could be a problem, but today most packing factories include heat treatment among their usual preserving systems.

Organoleptic analysis conducted at industrial scale has shown that the products prepared with reused brine, but following the traditional process in all other operations, had only very slightly higher scores (slightly lower acceptance) than olives without reused brines. In general, products prepared with active charcoal regenerated brines were always evaluated as slightly worse than those containing brines regenerated by ultrafiltration or than the control packed with fresh brine.

Brines regenerated by adsorption with activated charcoal and by ultrafiltration have also been successfully used in different proportions to prepare anchovy-stuffed olives. Organoleptic characteristics were also better for those fruits packed with brines treated by ultrafiltration (Rejano et al., 1995).

TABLE 11.6. Organoleptic Evaluation of Packed Products Reusing Regenerated Fermentation Brines in Different Proportions (Garrido Fernández et al., 1992. Reproduced with permission from *Grasas y Aceites*).

	Regenerated Brines		
	Usual Packing	Active Charcoal	Ultrafiltered
Without pasteurization (30%)[a]	1.8	2.1	2.1
Pasteurized (40%)[a]	1.7	2.4	1.9

[a]Percentages of reused brines in the final packing.

CONCLUSION

From the work conducted until now on green table olive brine regeneration, it can be stated that partial purification and reuse of fermentation green, Spanish-style table olive brines is feasible, and the necessary technology is ready. Implementation at industrial scale requires only small processing changes. Pasteurization must be applied to all products. This is not a drawback because such heat treatment is being adopted as habitual preservation system in the table olive packing industry.

ACKNOWLEDGEMENT

The authors thank the CICYT (Spanish Government) for financial support (ALI-97-0646) and to EU (FAIR CA-CT96-1118).

REFERENCES

Brenes Balbuena, M. and Garrido Fernández, A. 1988. Regeneración de salmueras de aceitunas verdes estilo español con carbón activo y tierras decolorantes. *Grasas y Aceites*, 39:96–101.

Brenes Balbuena, M., García García, P., and Garrido Fernández, A. 1986. Estudio comparativo de sistemas de conservación de aceitunas tipo negras. II. Efecto sobre las características del producto final. *Grasas y Aceites*, 37:301–306.

Brenes Balbuena, M., García García, P., and Garrido Fernández, A. 1988a. Regeneration of Spanish style green table olive brines by ultrafiltration. *Journal of Food Science*, 53:1733–1736.

Brenes Balbuena, M., García García, P., and Garrido Fernández, A. 1989. Influencias en el envasado de aceitunas verdes estilo español del reuso de salmueras regeneradas. *Grasas y Aceites*, 40:182–189.

Brenes Balbuena, M., Montaño Asquerino, A., and Garrido Fernández, A. 1990. Ultrafiltration of green table olive brines: Influence of some operating parameters and effect on polyphenol composition. *Journal of Food Science*, 55:214–217.

Brenes Balbuena, M., Sánchez, F., and Garrido, A. 1988b. Coagulación-filtración de salmueras de aceitunas verdes estilo español. *Grasas y Aceites*, 39:264–271.

Castro Gómez-Millán, A., Durán Quintana, M. C., García García, P., Garrido Fernández, A., González Cancho, F., Rejano Navarro, L., Sánchez Roldán, F., and Sánchez Tébar, J. C. 1981. *Reducción del volumen de vertidos en el proceso de elaboración de aceitunas verdes mediante reutilización de lejías, "cocidos" a baja concentración, y supresión de lavados*. Asamblea de Miembros del Instituto de la Grasa, Sevilla.

Fernández Díez, M. J., Castro Ramos, R. de, Garrido Fernández, A., Heredia Moreno, A., Gónzalez Cancho, F., Gónzalez Pellissó, F., Nosti Vega, M., Mínguez Mosquera, M. I., Rejano Navarro, L., Sánchez Roldán, F., Castro Gómez Millán, A. De, and García García, P. 1985. *Biotecnología de la Aceituna de Mesa*. Consejo Superior de Investigaciones Científicas, Madrid.

García García, P. and Garrido Fernández, A. 1984. Depuración parcial, por precipitación, de las lejías y aguas de lavado de la elaboración de aceitunas verdes estilo sevillano. *Grasas y Aceites,* 35:295–299.

García García, P., Brenes Balbuena, M., Vicente Fernández, J., and Garrido Fernández, A. 1990. Depuración de las aguas residuales de las plantas envasadoras de aceitunas verdes mediante tratamientos físico-químicos. *Grasas y Aceites,* 41:263–269.

Garrido Fernández, A. 1975. Tratamiento de las aguas residuales de la industria del aderezo. Métodos para su eliminación o reacondicionamiento para su posterior empleo. *Grasas y Aceites,* 26:237–244.

Garrido Fernández, A. 1994. Table olives in Spain: A current overview. Olivae, 50:21–27.

Garrido Fernández, A., Brenes Balbuena, M., and García García, P. 1992. Tratamiento de salmueras de fermentación de aceitunas verdes. *Grasas y Aceites,* 43:291–298.

Garrido, A., García, P., and Brenes, M. 1995. Olive fermentations. In *Biotechnology,* 2 ed., G. Reed and W. Nagodawithana (Eds.), pp. 593–627. VCH, Weinheim (Germany).

Rejano, L., Brenes, M., Sánchez, A. H., García, P., and Garrido, A. 1995. Brine recycling: Its application in canned anchovy-stuffed olives and olives packed in pouches. *Science des Aliments,* 15:541–550.

CHAPTER 12

Use of Trehalose for Osmotic Dehydration of Cod and Underutilized Fish Species

M. H. BRENNAN
T. R. GORMLEY

INTRODUCTION

TRADITIONALLY, osmotic dehydration of fish has involved soaking or coating fish with salt and/or sucrose (Collignan and Raoult-Wack, 1994). Salty foods are becoming less popular in the United Kingdom and Ireland as people try to reduce their salt intake for health reasons. In view of this trend, alternative solutes for the osmotic dehydration of fish were considered. The aim of the work reported here was to assess trehalose as a solute for the osmotic dehydration of fish.

Trehalose is a non-reducing disaccharide sugar. It is found in high concentrations in organisms that are able to withstand long periods of drought or sub-zero temperatures (Roser, 1991). Trehalose can function in cells as a substitute for water but, unlike water, does not evaporate or freeze. In the absence of trehalose, the cells and tissues become deformed and ruptured when water is lost from the organism. However, the structure of the trehalose molecule gives it a shape that resembles several water molecules superimposed on one another, and it is thought to fulfill the role of a barrier that protects the cell structure from deformation. Trehalose has many interesting properties (Roser, 1991) and so may have many potential food applications, particularly for dried or frozen foods. The reason it has not been exploited widely in the food industry is that it was very expensive to manufacture. However, a Japanese company has developed an enzymatic method for making food-grade trehalose from starch. The resulting trehalose is cheap enough to be used in foods. It is legally permitted for food use in Japan, whereas in Europe, British Sugar is currently seeking this permission, and it is anticipated that it will soon be

133

granted. Therefore, it is a pertinent time to be developing food applications for trehalose.

The quality of 23 underutilized fish species are being examined at The National Food Centre. These non-quota, underutilized fish species may have potential for commercialization as stocks of traditional fish species become depleted. Fifteen of the species were preferred to cod in taste panels of fillets, nuggets, or fish cakes. However, some of the species have high drip loss when thawed, and so it was hoped that osmotic dehydration could help to overcome this freezing damage.

EXPERIMENTAL

THE FISH

Three underutilized fish species, ling (*Molva molva*), Baird's smoothead (*Alepocephalus bairdii*), and orange roughy (*Hoplostethus atlanticus*), were selected for osmotic treatment on the basis of their availability, high taste panel acceptability, and inherent problems such as high drip loss on thawing. These fish were caught by personnel from the Fisheries Research Centre of the Marine Institute in two deep-water surveys of the eastern slopes of the Rockall Trough. All the underutilized species were frozen at sea and then stored at $-25°C$. Cod (*Gadus morhua*) was bought fresh from a local fish supplier, some fillets were frozen at $-35°C$ and stored at $-25°C$, others were stored for up to 2 days at 4°C.

EFFECT OF TREHALOSE CONCENTRATION AND SOAK TIME ON LING MASS

Thawed ling flesh was cut into cubes of approximately $2 \times 2 \times 2$ cm and weighed. The cubes were individually placed into jars of 40 ml of trehalose solution (0, 10, 30, 50, or 60% w/v) and soaked at room temperature (~20°C) and atmospheric pressure, without agitation. The cubes were removed after 1, 20, 25, 45, and 50 h, rinsed briefly in distilled water, patted dry on absorbent paper and weighed, and then returned to the trehalose solution. The experiment was repeated three times.

EFFECT OF FREEZING AND THAWING ON OSMOTIC DEHYDRATION OF COD

Cod was used for this experiment because fresh underutilized species were unavailable. Fresh and thawed cod fillets were cubed (~2 $\times$ 2 $\times$ 2 cm), and each piece (~10 g) was individually soaked in 40 ml of solution [0, 30, or 50% (w/v) trehalose). The soak was conducted at 4°C, to reduce microbial growth, under atmospheric pressure and without agitation. Samples were removed after 5 and 15 h and then briefly patted dry with absorbent paper. The water contents (by oven drying at 103°C) of unsoaked and soaked fresh and thawed cod were determined.

EFFECT OF SOAK TIME ON OSMOTIC DEHYDRATION OF BAIRD'S SMOOTHEAD AND COD

Baird's smoothead was used because it has a high drip loss (32% of frozen mass) on thawing; it was compared to cod, which has a low drip loss (1%). The thawed fish were cut into cubes (~2 × 2 × 2 cm) of about 10 g and individually placed in 40 ml of trehalose solution (50% w/v). The fish cubes were soaked at 4°C at atmospheric pressure for up to 7 h. After every hour, a cube of cod and of Baird's smoothead were removed from the cold room, and the remaining samples were swirled briefly by hand to mix. The fish cubes were patted briefly with absorbent paper and then weighed, and their water contents were determined by oven drying at 103°C. The soaking solutions were filtered, and their water content and nitrogen content (Leco FP328) were determined.

EFFECT OF OSMOTIC DEHYDRATION ON THE PHYSICAL PROPERTIES OF ORANGE ROUGHY

Orange roughy was chosen because it was the most preferred of all the underutilized fish species tested at the National Food Centre and because it has high drip loss on thawing.

Orange roughy fillet was thawed and the skin was removed; the water content of a sample (~10 g) was determined by oven drying at 103°C. Two similar shaped pieces (10 × 5 × 2 cm) were weighed, and one was soaked at 4°C for 1 h in 250 ml of 50% (w/v) trehalose solution, under atmospheric pressure with no agitation. The other piece (the control) was stored at 4°C without soaking. The osmotically treated sample was briefly patted dry with absorbent paper and then a small sample was removed to determine its water content. The control and osmotically treated samples were weighed, frozen at −35°C, stored at −25°C overnight, and then thawed. The samples were reweighed to determine drip loss on thawing. The color of the flesh was measured (Minolta chromameter, Hunter Lab color scale). The samples were cooked in a water bath at 80°C in sealed bags. The fish were cooled and reweighed, their color was remeasured, and a sample was removed to determine water content. Finally, the shear value of 80 g of cooled, cooked fish was measured with a Kramer style shear press (T-2000 Texture System) with a standard shear compression cell (model CS-1). The experiment was repeated with fillet pieces of ~11 × 3 × 2 cm by using the same method except that they were frozen at −25°C, instead of −35°C, to increase the potential damage caused by freezing.

RESULTS AND DISCUSSION

EFFECT OF TREHALOSE CONCENTRATION AND SOAK TIME ON LING MASS

All the samples gained mass during the first ~24 h of soaking (Table 12.1). The fish soaked in the highest trehalose solution (60% w/v) gained the most mass. This result suggests that the fish took up trehalose but does not clarify whether the fish were dehydrated by osmosis. After ~25 h, the fish soaked in 10 and 30% trehalose started to lose mass (Table 12.1); it is speculated that this might be associated with a degradation of the fish and that the higher trehalose content of the fish soaked in 50 and 60% trehalose solutions gave protection against this. However, this observation was not investigated further.

The fish sample soaked in water (i.e., 0% trehalose) increased in mass (Table 12.1), which must be due to water uptake. The amount of water taken up during the soak might be related to the amount of water previously lost from the fish due to freezing and thawing.

EFFECT OF FREEZING AND THAWING ON OSMOTIC DEHYDRATION OF COD

The results show that soaking both fresh and thawed cod in trehalose solution (30 or 50% w/v) reduced the water content of the fish. The initial (0 h) water content of thawed fish was lower than that of the fresh fish but after a

TABLE 12.1. Fish Mass Increase[a] as a Percentage of the Mass Prior to Soaking.

Soak Time (h)	Trehalose Concentration of Soak Solution (% w/v)				
	0	10	30	50	60
1	2.8	3.2	11.0	11.1	13.1
20	5.4	9.9	24.3	25.3	37.9
25	4.3	8.5	22.9	25.8	37.5
45	2.0	—	—	27.4	37.9
50	4.4	27.7–28.9	10.5–22.0	26.0	35.3

[a]Mean from four replicates.

TABLE 12.2. The Water Content[a] of Fresh and Thawed
Cod Soaked in Trehalose Solution.

Soak Time (h)	Fresh Cod			Thawed Cod		
	0%[b]	30%[b]	50%[b]	0%[b]	0%[b]	50%[b]
0	8	8	8	78		78
5	3.8	3.8	3.8	.0	8.0	.0
1	8	7	6	82		63
5	3.7	4.6	8.8	.0	3.4	.8
	8	7	6	82		62
	5.4	1.5	3.4	.6	0.3	.8

[a]% w/w; mean from two samples determined by oven drying (103°C).
[b]Trehalose concentration (% w/v) of soak solution.

5-h soak in 30% or 15 h in 50% (w/v) trehalose solution the water contents were similar for fresh and thawed cod samples (Table 12.2). It was shown that the mass of water removed by osmotic dehydration was influenced by pre-freezing and thawing, but the final water content of the fish was influenced by the trehalose concentration in the soak solution. The water content of the fresh fish soaked in water increased only slightly, whereas that of the thawed fish increased to a greater extent (Table 12.2).

EFFECT OF SOAK TIME ON OSMOTIC DEHYDRATION OF BAIRD'S SMOOTHEAD AND COD

The results show that the soaking treatment increased the fish mass (Table 12.3) but reduced the water content (Table 12.4). Therefore, the mass increase must have been due to uptake of trehalose.

The water content of the fish decreased from 0 to 5 h soak for cod and from 0 to 4 h for Baird's smoothead (Table 12.4). The highest change in water con-

TABLE 12.3. Mass Increase[a] for Fish Soaked
in 50% (w/v) Trehalose Solution.

Soak Time (h)	Cod	Baird's Smoothead
1	13.6%	22.0%
2	10.0%	14.1%
3	12.4%	26.8%
4	19.7%	35.5%
5	21.8%	19.7%
6	15.3%	13.3%
7	13.5%	25.1%

[a]As percentage of mass before soak.

TABLE 12.4. **Water Content**[a] **of Fish Soaked in 50% (w/v) Trehalose Solution.**

Soak Time (h)	Cod	Baird's Smoothead
0	79.7	83.1
1	72.6	72.8
2	72.6	70.5
3	72.4	69.6
4	68.6	68.5
5	67.6	70.8
6	75.6	72.0
7	684	62.1

[a]As percentage of wet weight.

tent per hour for the fish and soaking solutions was in the first hour (Tables 12.4 and 12.5).

The pattern of the results changed after 4–5 h of soaking (Tables 12.4 and 12.5); it was hypothesized that this might have been due to degradation of the fish, which could be monitored by determining the nitrogen concentration in the soak solutions. However, nitrogen was found in the soak solution after 1 h, and the maximum concentration found was after 5 h soak for cod and 4 h soak for Baird's smoothead, but it decreased thereafter (Table 12.6). Further expla-

TABLE 12.5. **Water Content**[a] **(as Percentage of Wet Weight) of Filtered Soaking Solutions.**

Soak Time (h)	Baird's Smoothead (%)	Cod (%)
0	67.3	6 7.3
1	69.9	7 0.1
2	70.8	7 0.3
3	70.5	7 0.7
4	71.0	7 1.1
5	71.1	7 1.9
6	71.4	7 1.5
7	70.5	7 1.0

TABLE 12.6. Nitrogen Concentration[a]
of Filtered Soaking Solutions.

Soak Time (h)	Cod	Baird's Smoothead
1	0.34	0.17
4	0.57	0.62
5	0.85	0.58
6	0.69	0.37
7	0.57	0.31

[a]g/L, determined for 2 ml of solution.

nations have not been proposed or investigated. It was concluded that a soak time of 1 h was suitable for subsequent osmotic treatments.

EFFECT OF OSMOTIC DEHYDRATION ON THE PHYSICAL PROPERTIES OF ORANGE ROUGHY

The results in Tables 12.7 and 12.8 show that the osmotic treatment increased fish mass (due to trehalose uptake) and reduced water content but did not reduce drip loss on thawing or cooking.

The osmotic treatment had little effect on the whiteness of raw or cooked fish, but it decreased shear value considerably (Tables 12.9 and 12.10). This is of potential benefit as toughening during freezing and thawing can lead to adverse consumer acceptability.

TABLE 12.7. Mass[a] and Water Content for Experiment 1
(Refrozen at −35°C).

| Process Step | Osmotically Treated Fish | | Control Fish | |
	Mass (g)	Water Content (%)	Mass (g)	Water Content (%)
Frozen	154	80	15	80
After thawing	136	77	4	77
	140	68	13	n/a
After soaking	133	66	6	76
	91	51	n/a	67
After rethawing			13	
			0	
After cooking			93	

[a]Samples were removed for water content determination after some of the process steps; however, the masses have been calculated/extrapolated so that this can be ignored.
n/a Not applicable.

TABLE 12.8. Mass[a] and Water Content for Experiment 1
(Refrozen at −35°C).

Process Step	Osmotically Treated Fish		Control Fish	
	Mass (g)	Water Content (%)	Mass (g)	Water Content (%)
Frozen	98	80	98	80
After thawing	81	76	81	76
	83	70	n/a	n/a
After soaking	79	69	77	74
	52	52	55	64
After rethawing				
After cooking				

[a]Samples were removed for water content determination after some of the process steps; however, the masses have been calculated/extrapolated so that this can be ignored.
n/a Not applicable.

TABLE 12.9. Effect of Osmotic Treatment on Fish
Color and Texture for Experiment 1.

Measurement	Osmotically Treated Fish	Control Fish
Whiteness[a] - raw	70	71
Whiteness[a] - cooked	74	77
Shear value (N/80 g[b])	1300	1430
Shear value N/g dry matter[c]	33	53

[a]Hunter L value.
[b]Wet weight. Note that shear value/g is non-linear.
[c]Calculated as: shear value (N/80 g)/dry matter mass in 80 g wet weight.
[d]Calculated as: shear value (N/40 g)/dry matter mass in 40 g wet weight.

TABLE 12.10. Effect of Osmotic Treatment on Fish
Color and Texture for Experiment 2.

Measurement	Osmotically Treated Fish	Control Fish
Whiteness[a] - raw	66	71
Whiteness[a] - cooked	80	81
Shear value (N/40 g[b])	1190	1550
Shear value N/g dry matter[d]	69	99

[a]Hunter L value.
[b]Wet weight. Note that shear value/g is non-linear.
[c]Calculated as: shear value (N/80 g)/dry matter mass in 80 g wet weight.
[d]Calculated as: shear value (N/40 g)/dry matter mass in 40 g wet weight.

CONCLUSION

The osmotic treatment of fish using trehalose solution had two main effects: to reduce water content and to incorporate trehalose into the fish. These changes did not result in a reduction of liquid loss on subsequent thawing or on cooking, but they reduced the toughness of cooked fish.

ACKNOWLEDGEMENTS

This research was funded in part under the Marine Research Measure of the Operational Programme for Fisheries. This program is administered by the Marine Institute of Ireland and is financed in part by the E.U.'s European Regional Development Fund.

REFERENCES

Collignan, A. and Raoult-Wack, A.-L. 1994. Dewatering and salting of cod by immersion in concentrated sugar/salt solutions. *Lebensmittel-Wissenschaft & Technologie,* 27:259–264.

Roser, B. 1991. Trehalose, a new approach to premium dried foods. *Trends in Food Science and Technology,* 2:166–169.

VACUUM SALTING PROCESSES

CHAPTER 13

Cheese Salting by Vacuum Impregnation

C. GONZÁLEZ-MARTÍNEZ
M. PAVIA
A. CHIRALT
V. FERRAGUT
P. FITO
B. GUAMIS

INTRODUCTION

SALTING of cheese in brine immersion (BI) has been described as a diffusion process, but capillary mechanisms may also play an important role in the salt uptake (Geurts et al., 1980). The capillary entry of the external brine in the curd pores can contribute to the total salt gain to quite an extent when subatmospheric pressure is applied to the salting tank. Capillary penetration in a pore occurs coupled with the compression of the occluded internal gas. The volume fraction of the sample penetrated by the external liquid (X_c) is a function of the capillary pressure (p_c), the system pressure (p), and the effective porosity of the product (ϵ) (Fito and Pastor, 1994). According to Equation (1), the lower the system pressure (p), the greater the capillary penetration.

$$X_c = \epsilon \cdot \left(\frac{p_c}{p + p_c} \right) \qquad (1)$$

When atmospheric pressure (p_2) is restored in a tank containing a porous product immersed in a liquid phase, initially under vacuum conditions (p_1), hydrodynamic mechanisms (HDM) act, in line with the compression of product residual gas, leading to a great external liquid penetration into the pores. This process has been described as vacuum impregnation (VI) (Salvatori et al., 1998). The volume fraction of the sample penetrated by the external liquid (X) has been modeled by Equations (2) and (3) (Fito and Pastor, 1994). Changes in the pressure system can also promote sample deformations, cou-

145

pled with sample impregnation, which will modify the initial sample porosity (Fito et al., 1996).

$$X = \epsilon \cdot \left(1 - \frac{1}{r}\right) \tag{2}$$

$$r = \frac{p_2}{p_1} + \frac{p_c}{p_1} \tag{3}$$

Enzyme-coagulated pressed curd shows a porous structure that is susceptible to impregnation by HDM or VI with concentrated brine, thus contributing to accelerate the slow diffusional salt gain. In previous studies, cheese salting by applying vacuum during brining has been described as reducing salting time, promoting a flatter initial salt concentration profile, and a more closed structure (Chiralt and Fito, 1997; Andrés et al., 1997). Nevertheless, differences between BI and VI cheese status could affect their ripening behavior.

Vacuum treatments have also been applied afterward, before or during pressing in cheese making of large blocks of cheddar cheese (Irvine and Burnett 1962; Reinbold et al., 1993). Vacuum application promotes the expulsion of the entrapped air and whey in the curds, which causes different effects in the product. Removal of air helped to reduce the size of mechanical openings in cheese (Czulak et al., 1962; Irvine and Burnett, 1962), whereas the release of whey contributes to reduce the water content, increasing the evenness of moisture distribution (Reinbold et al., 1993).

The aim of this work is to analyze the effect of VI in salting kinetics of Manchego cheese as a function of the cheese-making variables affecting VI effectiveness through curd porosity, such as pressing conditions and time between pressing and salting steps. Ripening behavior of BI and VI cheeses were compared in some ripening indexes, mechanical and structural properties, and drying pathway.

INFLUENCE OF DIFFERENT FACTORS ON THE VI EFFECTIVENESS IN CHEESE SALTING

The VI effectiveness depends on the coupling of two phenomena: liquid penetration and solid matrix deformation during pressure changes (Fito et al., 1996). So, the impregnation and deformation levels of the product at mechanical equilibrium will be affected by size and shape of the pores, mechanical response of solid matrix, and viscosity of the external solution; these variables affect the penetration-deformation characteristic times (Fito et al., 1996). During cheese making, curd processing and pressing conditions greatly affect final cheese porosity. So, different pressing conditions will lead to different yield of salting by VI. Figure 13.1 shows the salt mass fraction reached in cheese liquid phase (water plus salt) for pieces obtained in different pressing conditions (time, pressure, and kind of press). Salt levels are in the usual range for this kind of cheese, and times required to reach this concentration are lower in all cases when vacuum is applied for a determined period. Nevertheless, a more notable difference can be observed when applying only 4-h pressing. No significant differences were appreciated between curds pressed for 17 and 7 h, respectively. Differences in salt penetration in the internal zone of the cheese (whole piece less an external part of 1.5 cm thickness) were only relevant for 4-h pressed curds. Too intense pressing conditions seem to reduce greatly curd porosity and, therefore, the HDM action, thus limiting the VI effectiveness.

Another aspect related with pressing effect concerns the heterogeneous pore distribution as a function of the distance to press plug in the mold. Because of the viscoelastic response of curd during dynamic deformation throughout pressing, a heterogeneous pressure distribution occurs in the mold, resulting in a different curd compactness; the greater the distance to plug, the lesser the structure compactness (González et al., 1999). This implies that when porosity plays an important role in salt uptake, such as VI brining, a faster salt entry will occur in the bottom part (as related to plug position) of the cheese.

Figure 13.2 shows the salt concentration profile in the bottom and top parts of a cheese salted for 17 h, as a function of the distance to the sample salting surface. Lateral surface of the cylindrical sample was covered with a Teflon film to avoid radial mass transfer. The profiles were obtained by sampling and analyzing 1.5-mm-thick slices taken from the interface to the sample center (Chiralt et al., 1999). The curves in Figure 13.2 show the different profile obtained in BI and VI cheeses. Although no differences between the bottom and

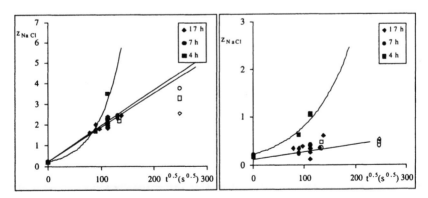

Figure 13.1 Mean (left) and internal (right) salt concentration in cheese liquid phase salted by BI (empty symbols) and VI (full symbols), pressed in different conditions (17 h and 7 h with vertical press and 4 h with horizontal press).

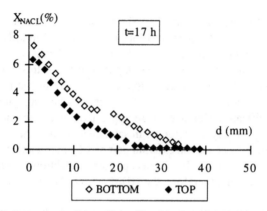

Figure 13.2 Salt concentration profiles achieved in the top and bottom parts of the cheese.

top profiles were observed in BI, a greater development of salt profile in the bottom part of VI pieces can be seen.

The last results suggest that a good control of pressing conditions needs to be applied to ensure regular salt uptake if VI is used to reduce salting time. In this sense, the horizontal press would not be recommendable because not all the cheeses receive pressure directly from the plug, but a pressure drop occurs through the cheeses coupled in series because of viscoelastic behavior of curd (Figure 13.3).

Another factor affecting feasibility of VI brining is the time between pressing and salting process steps. Curd porosity reduces 24 h after pressing, and no HDM salt penetration has been observed in this case (González et al, 1999).

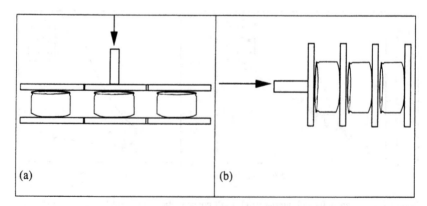

Figure 13.3 Scheme of vertical (a) and (b) horizontal press.

TABLE 13.1. Salting Times and Water and Salt Mass
Fractions Obtained in BI and VI Treatments.

Treatment	t_1 (h)	t_2 (h)	X_{NaCl} (g/100 g)	X_w (g/100 g)	$X_R{}^a$ (g/100 g d.m.)
BI	0	17	2.03 ± 0.09	47.2 ± 0.7	5.84
VI	1.75	1.75	1.98 ± 0.03	48.7 ± 0.1	5.37

[a]Salt mass fraction in the cheese rind (0.5–1 cm thick).

EFFECT OF VI ON SALTING TIME AND INITIAL SALT DISTRIBUTION

As commented on above, when moderate pressing conditions are applied, a great reduction of salting time has been observed by applying vacuum for a period in the salting tank. Table 13.1 shows the salting times required to reach similar salt content in a determined cheese batch (pressed for 1 h at 0.008 kg/cm^2 and for 3 h at 0.016 kg/cm^2) in BI and VI processes. The salting time needed to obtain the same overall salt content by BI is five times shorter than when vacuum was applied. The salt distribution in a representative cheese sector immediately after salting can be observed in Figure 13.4, in salt iso-concentration curves. In ordinates, the distance to the cheese equatorial plane was plotted, and in X-axis the radial distance. The more external part (rind: 0.5–1 cm thick) is not plotted in Figure 13.4, and its mean salt concentration value is given in Table 13.1. In Figure 13.4 concentration was referred to dry matter to make comparisons independent of water distribution.

The deeper penetration of salt into the cheese in the VI process due to HDM action is evidenced in Figure 13.4. Greater accumulation of salt in the rind (Table 13.1) and in the more external zone (approximately 1.5 cm around the

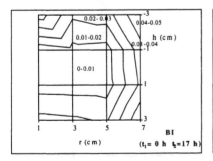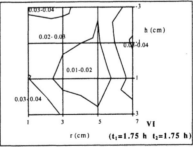

Figure 13.4 Salt distribution (dry basis ratio) immediately after salting process achieved in BI and VI treatments.

rind) is observed for BI treatment, leaving an internal zone with practically no salt concentration.

INFLUENCE OF VI ON CHEESE RIPENING

RIPENING INDEXES

The different salt profile in BI and VI cheeses at the first ripening period could cause different development of biochemical reactions, depending on the cheese zone. However, different studies (Guamis et al., 1997; Andrés et al., 1996; Pavía, et al., 1999a) did not show important differences in conventional indexes such as soluble nitrogen fractions or residual lactose. Figure 13.5 shows the development of total soluble nitrogen (SN) vs. time for different cheese zones [internal, intermediate, and rind, according to previous article, Guamis et al. (1997)] and salting treatment. A very similar pathway was observed for BI and VI cheeses for all zones, although in the rind a slightly greater SN content in VI pieces was detected, as expected from the lesser salt levels reached in this zone for this treatment. VI did not cause significant differences in residual lactose concentrations in the internal and intermediate areas, yet it did in the rind (Table 13.2). VI rind shows a lower residual lactose amount than BI rind. This may be due to the higher salt-in-moisture value that rind of BI cheeses has after salting, which is never reached by VI pieces. Differences observed in the internal and intermediate areas are negligible but agree with the higher initial salt levels in these zones in VI pieces.

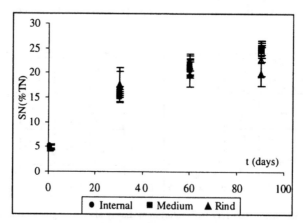

Figure 13.5 Soluble nitrogen in the internal, intermediate, and rind part of the cheeses salted by BI (full symbols) and VI (empty symbols) throughout ripening.

TABLE 13.2. Mean and Standard Deviation Values of Residual Lactose Concentrations (%) Immediately after Salting.

	BI		VI	
Rind	0.501	0.015	0.465	0.054
Medium	0.052	0.016	0.067	0.020
Internal	0.040	0.020	0.060	0.021

Adapted from Pavia et al., 1999b.

MECHANICAL AND STRUCTURAL PROPERTIES

Mechanical properties of BI and VI cheeses at 7 and 20°C have been compared at different times of ripening. Values of the true stress (σ_F) and strain (ϵ_F) at the fracture point obtained from a uniaxial compression test, conducted in conditions previously described (Pavía et al., 1999b), are given in Table 13.3. As expected, cheeses became shorter and firmer to the fracture in line with ripening. In determined conditions, the fracture resistance of VI cheeses is higher than that of BI samples, especially at the lower temperature and for the longer ripening times. No notable differences are appreciated in cheese shortness because of the salting procedure. Differences in resistance could be explained in the differences in protein matrix compactness induced by vacuum treatment, as commented on below.

From a macrostructural point of view, VI implies a reduction and final disappearance of the mechanical eyes, which can be a positive or negative factor, depending on consumer expectation. Microstructural observations were conducted in samples stained with Nile Blue 0.2% with a CLSM at every stage of ripening. These show differences in the structure of the protein matrix between BI and VI.

TABLE 13.3. Stress and Strain at the Fracture Point of Cheeses Obtained from Uniaxial Compression Test at 7 and 20°C for Different Times of Ripening.

Time	Treatment	Fracture Stress (N/m²) 10^{-4}		Fracture Strain	
		7°C	20°C	7°C	20°C
Day 1	BI	7.5 ± 1.3	3.7 ± 0.3	0.75 ± 0.09	0.78 ± 0.05
	VI	8.7 ± 2.5	5.1 ± 0.3	0.69 ± 0.13	0.73 ± 0.03
Day 30	BI	6.2 ± 0.9	3.2 ± 0.5	0.42 ± 0.04	0.42 ± 0.09
	VI	6.7 ± 1.2	3.3 ± 0.1	0.44 ± 0.05	0.49 ± 0.06
Day 60	BI	7.3 ± 1.1	4.4 ± 0.1	0.32 ± 0.05	0.38 ± 0.03
	VI	10.6 ± 1.8	4.9 ± 0.1	0.34 ± 0.02	0.4 ± 0.03
Day 90	BI	11.4 ± 2.3	4.5 ± 0.4	0.27 ± 0.01	0.34 ± 0.02
	VI	17.5 ± 3.5	5.1 ± 0.9	0.28 ± 0.02	0.41 ± 0.03

Figure 13.6 shows the micrographs of protein matrix of both kinds of cheeses at 60 days ripening. Vacuum impregnated cheeses show a more compact protein network, with small pores and small and regular fat globules homogeneously distributed in the continuous matrix. On the other hand, conventional brine-immersed cheeses show less homogeneous pore distribution in the protein matrix, including matrix holes containing fat globules.

Because there is a close relationship between cheese microstructure and cheese rheology (Stanley and Emmons, 1977; Yiu, 1985), the observed microstructural differences can explain the differences in mechanical behavior commented on above.

DRYING BEHAVIOR

Drying behavior of Manchego type cheese has been modeled as a function of salt content and salting method, both inducing a different initial salt profile and thus different water potential gradients in the cheese (González et al., 1998). Figure 13.7 shows the drying curves of BI and VI cheeses containing the same salt content in the usual range (1.8–2%). Two different steps were observed in drying curves in both cases. In the first, up to about 20 days ripening, a falling rate period was observed, which has been modeled through a diffusional approach, obtaining a water effective diffusivity (D_e) for each kind of cheese and a kinetic constant to describe the cheese height reduction in this period. Significant differences in the values of D_e were observed for BI and VI cheeses, these being higher for BI pieces. This implies a faster drying of BI pieces in the first period that was explained in the contribution, as drying driving force, of the abrupt water chemical potential gradients inferred during

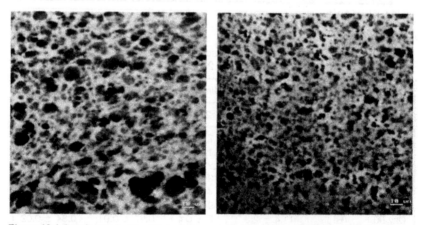

Figure 13.6 Protein matrix (in white) of cheeses salted by BI (a) and VI (b) at 60 days ripening.

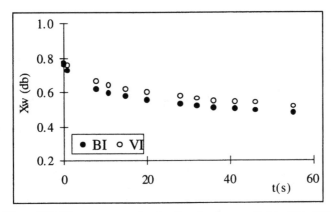

Figure 13.7 Water content vs. ripening time for cheeses salted by BI and VI.

salting. In the second step, an almost constant drying rate was deduced from drying curve, with no differences among cheeses for kinetic parameters.

REFERENCES

Andrés, A., Martínez-Navarrete, N., Panizzolo, L., and Chiralt, A. 1996. Caracterización físico-química y reológica de quesos tipo Manchego salados por inmersión y por impregnación a vacío. En *Equipos y procesos para la industria alimentaria. Tomo II: Análisis cinético, termodinámico y estructural de los cambios producidos durante el procesamiento de alimentos*, E. Ortega, E. Parada y P. Fito (Eds.). Servicio de Publicaciones de la Universidad Politécnica de Valencia, pp. 549–561.

Andrés, A., Panizzolo, L., Camacho, M. M., Chiralt, A., and Fito, P. 1997. Distribution of salt in manchego type cheese after brining. In *Engineering & Food at ICEF 7, De.,* R. Jowitt (Ed.). Sheffield: Academic Press, pp. A133–A136.

Chiralt, A. and Fito, P. 1997. Salting of Manchego cheese by vacuum impregnation. In *Food Engineering 2000*, Fito, P., Barbosa, G., and Ortega, E. (Eds.). New York: Chapman and Hall, pp. 119–215.

Chiralt, A., Fito, P., Gonzalez-Martínez, C., and Andrés, A. 1999. Salt transport in pressed curd as affected by brine vacuum impregnation. *Proceedings of the 6th Conference of Food Engineering*, pp. 106–111.

Czulak, J., Freeman, N. H., and Hammond, L. A. 1962. Close texture in cheddar cheese by vacuum pressing. *The Australian Journal of Dairy Technology,* 17:22–25.

Fito, P. and Pastor, R. 1994. Non-difusional mechanism occurring during vacuum osmotic dehydration. *Journal of Food Engineering,* 21:513–519.

Fito, P., Andrés, A., Chiralt, A., and Pardo, P. 1996. Coupling of hydrodinamic mechanism and deformation relaxation phenomena during vacuum treatments in solid porous-liquid systems. *Journal of Food Engineering,* 27:229–240.

Geurts, T. G., Walstra, P., and Mulder, P. 1980. Transport of salt and water during salting of cheese. 2. Quantities of salt taken up and of moisture lost. *Neth. Milk Dairy J.,* 34:229–254.

González, C., Martínez, N., Chiralt, A., and Fito, P. 1998. Drying behaviour of Manchego type cheese throughout ripening. *Proceeding of the 11th International Drying Symposium,* Greece, vol. B, pp. 1251–1258.

González, C., Fuentes, C., Andrés, A., Chiralt, A., and Fito, P. 1999. Effectiveness of vacuum impregnation brining of manchego type curd. *International Dairy Journal,* 9:143–148.

Guamis, B., Trujillo, J. A., Ferragut, V., Chiralt, A., Andrés, A., and Fito, P. 1997. Ripening control of Manchego type ewe's cheese salted by brine vacuum impregnation. *International Dairy Journal,* 7:185–192.

Irvine, R. O. and Burnett, K. A. 1962. Effects of vacuum treating on the textured qualities of Cheddar cheese. *Canadian Dairy and Ice Cream Journal,* 41:24–28.

Pavia, M., Guamis, B., and Ferragut, V. (1999a). Effects of ripening time and salting method on glicolisis in Manchego-type cheeese. *Milchwissenst,* 54(7):379–381.

Pavia, M., Guamis, B., Trujillo, A. J., Capellas, M., and Ferragut, V. 1999b. Changes in microstructural, textural and colour characteristics during ripening of Manchego-type cheese salted by brine vacuum impregnation. *International Dairy Journal,* 9(2):91–98.

Reinbold, R. S., Hansen, C. L., Gale, C. M., and Ernstrom, C. A. 1993. Pressure and temperature during vacuum treatments of 290-kilogram stirred-curd cheddar cheese blocks. *Journal Dairy Science,* 76:909–913.

Salvatori, D., Andrés, A., Albors, A., Chiralt, A., and Fito, P. 1998. Structural and compositional profiles in osmotically dehydrated apple. *Journal of Food Science,* 63:606–610.

Standley, D. W. and Emmons, D. B. 1977. Cheddar cheese made from bovine pepsin II texture-microstructure-relationships. *Can. Inst. Food csi. Technol. J.,* 10:18–84.

Yiu, S. H. (1985). A fluorescent microscopy study of cheese. *Food Microstructure,* 4:99–106.

CHAPTER 14

Salting Time Reduction of Spanish Hams by Brine Immersion

J. M. BARAT
R. GRAU
A. MONTERO
A. CHIRALT
P. FITO

INTRODUCTION

S PANISH dry-cured ham is a nonsmoked meat product manufactured according to ancient traditions: stabilization through salt diffusion, a decrease in water activity, and development of the typical flavor throughout a long period of maturation. Proteolysis and lipolysis increase during the aging of country-style hams (Ockerman et al., 1963), Parma hams, French hams, and Spanish hams (García-Regueiro and Diaz, 1989). The phenomena undoubtedly contribute to flavor development.

The dry-curing process is the oldest one and uses a mixture of curing ingredients, mainly salt, nitrate and/or nitrite, and sugars. Generally, the dry-curing is applied without any added water. Consequently, the curing agents are solubilized in the original liquid present in the meat, and they mainly penetrate by diffusion. The dry-curing process is applied to pieces and minced meats (Flores, 1996).

Processes can vary in small details, depending on the traditions of each area of production, but they always have three fundamental stages:

(1) Salting: Hams are completely covered with a mixture of the curing ingredients and salt and put in a cold storage room (1–3.3°C).
(2) Post-salting: During this stage the salt equalization takes place. The salted hams are kept at low temperatures (approximately 3.3°C) for approximately 30 days.
(3) Dry-maturation: During this stage proteolysis and lipolysis take place. These reactions generate the desirable sensory characteristic for which dry-cured ham is so appreciated.

155

The salting stage is the shortest stage in the process (usually 1–1.5 days kg^{-1}, more or less 9–14 days), and it has a great influence on the quality of the final product. During this stage, the salt uptake takes place, but salt concentration is not the same in all ham muscle. Salt diffusion is temperature dependent, but only a narrow temperature range can be used during the salting stage of the dry-cured ham process, because of the risk of microorganism growth (Pérez-Alvarez et al., 1997).

The traditional salting stage implies the use of dry salt, and during the process, the solid salt is partially dissolved and drained because of the water flux from the meat due to osmotic and diffusional mechanisms. The formed brine, with a high organic content, is discharged to the sewage and is enormously difficult to purify because of the NaCl presence.

Over the last few years, new technologies have been incorporated into the dry-cured ham process to improve the traditional homemade method. One of the affected operations is the salting stage. It has been seen that salting can be conducted more quickly in porous products (such as cheese, meat, etc.) introducing brine into the pores by means of VI processes (Fito et al., 1994; Chiralt and Fito, 1997; Iriarte et al., 1993). Vacuum brine impregnation (VBI) consists of two steps: first, a vacuum pressure is applied and maintained in the brine tank during a time t_1 and second, the atmospheric pressure is restored while the hams remain immersed for a time t_2. During the vacuum period, the gas occluded in the porous structure of the ham expands and partially flows out, allowing a more intense capillary penetration. In the second period, the restoring of the atmospheric pressure promotes the residual gas compression and the external brine penetration by a hydrodynamic mechanism (HDM) (Fito et al., 1996; Fito, 1994; Fito et al, 1994), thus allowing a faster salting process.

After salting is completed, the excess salt is washed off, and meat is placed under refrigeration for 20–40 days until salt equalization. The objective of this post-salting phase is to achieve a complete and homogeneous salt distribution throughout the piece of meat, whereas the risk of microorganism deterioration is avoided by keeping the system at refrigeration temperatures (3–5°C). The next step (dry-maturation or ripening period) involves time-temperature combinations. The homogenized pieces are placed in natural or air-conditioned drying chambers where the relative humidity is usually varied between 90 and 70%, and the temperature range is from 5 to 26°C (Tapiador Farelo, 1989; Bañon et. al., 1997; Gelabert et. al., 1998). The ripening period varies with the type of ham from a minimum of 6 months (rapid process) and up to 24 months (slow process).

The aim of this work was the study of brine salting (with and without vacuum impregnation), as an alternative to pile salting in ham processing, with the benefits of a faster process and the possibility of brine reuse, with the subsequent reduction of environmental impact. The influence of salting process over the following stages was also studied.

MATERIALS AND METHODS

RAW MATERIAL

Experiments were accomplished with fresh hams selected in such a way that the two hams from the same pig were used in two sets of experiments and compared to minimize the variability due to the raw material. Fresh ham weight was 10.23 ± 0.827 kg.

Two ham batches were processed (30 hams each batch) and compared; dry salting (DS I) and brine vacuum impregnation (BVI) were used in the first batch, and dry salting (DS II) and brine immersion (BI) were used in the second one.

SALTING PROCEDURE

- Dry salting: This is the traditional method used as reference. Salting was conducted in containers using dry salt. According to ham weight, two different salting times were chosen, 9 and 10 days for the first and second batch, respectively, while salting temperature was kept at 3°C.
- Brine salting: Hams were salted with 24% (w/w) brine in a tank with controlled temperature and pressure. Salting temperature was 3°C, and 50 mbar of pressure was applied during the vacuum period (20 h). The equipment was designed and built by the DTA-UPV and METALQUIMIA, S.A. (Figure 14.1).

Salting time needed was calculated on the basis of a previous study in which salting kinetic was determined for the two brine salting methods (Barat et al., 1998). In that work, the experimental results were fitted to Equation (1) on the basis of the salt concentration of the product liquid phase (z^{NaCl}). The parameters of the fitted equation can be observed in Table 14.1.

$$z^{NaCl} = k_1 \cdot t^{0.5} + k_2 \qquad (1)$$

It was considered that the goal of the salting process was a z^{NaCl} value of 0.042. This value was obtained by taking into account that many of the ham manufacturers try to get a NaCl content after the salting process of 3% and that the average moisture of the salted hams was found to be 68.4%. To obtain the mentioned salt content ($z^{NaCl} \sim 0.042$), the required time in BI and BVI would be 6.9 and 4.2 days, respectively, according to the linear predictions. Because salted hams were going to be processed to reach the final product (9.5 more months), it was important to ensure a secure salt content to obtain the microbial stabilization. That was the reason why the salting time chosen was 5.5 days for BVI (the first 20 h under vacuum conditions) and 9 days for BI, clearly in excess of the predicted ones.

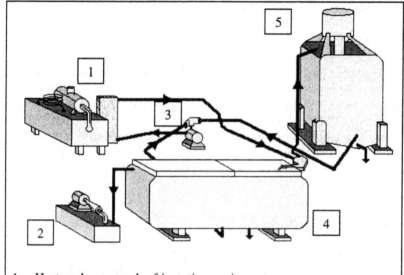

1. Heat exchanger and refrigeration equipment.
2. Vacuum pump.
3. Brine pump.
4. Treatment tank where hams were submerged in the brine. This tank is prepared to work at subatmospheric pressures.
5. Brine preparation tank.

Figure 14.1 Outline of the salting equipment.

POST-SALTING AND DRY-MATURATION CONDITIONS

These stages were conducted in a ham-manufacturing industry using industrial equipment. Dry and brine salted hams followed the same stages and temperature-relative humidity conditions as the commercial hams (dry salted).

Process conditions in the post-salting, drying, and ripening stages are shown in Table 14.2.

TABLE 14.1. Kinetic Parameters for Brine Salting Experiments.

	k_1 ($h^{0.5}$)	k_2
BVI	0.0042	$3*10^{-5}$
BI	0.0032	$8*10^{-4}$

TABLE 14.2. Processing Conditions.

Stage	Salting Method	Time (days)	$T°$ (°C)	H (%)
Posting-Salting	BVI	49	2–3.5	80–85
	BI	50	2–3.5	80–85
	DS	50	2–3.5	80–85
Drying				
Cycle 1	BVI / BI / DS	15	6–8	80–85
Cycle 2	BVI / BI / DS	25	10–12	75–80
Cycle 3	BVI / BI / DS	20	16–18	75–80
Cycle 4	BVI / BI / DS	20	20–22	75–80
Cycle 5	BVI / BI / DS	25	22–24	75–80
Cycle 6	BVI / BI / DS	45	24–26	75–80
Ripening	BVI	32	28–30	75–80
	BI	32	28–30	75–80
	DS	25	28–30	75–80

SAMPLING

In all cases, initial and final weight of ham was measured. Salt and moisture concentration analyses were conducted by using four samples; three of them were representative of the widest section of ham (A, B, and C; Figure 14.2), and the fourth was the homogenized remainder (R). Each sample was thoroughly homogenized before the moisture and sodium chloride content determination.

In all cases, three hams were taken at each sampling time, batch and salting method.

After the salting period, hams were taken out, weighed, and placed for 24 h in a chamber. Ambient temperature and relative humidity were controlled

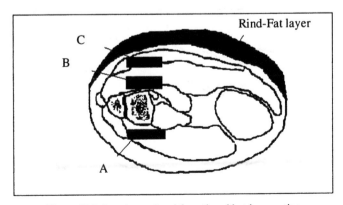

Figure 14.2 Samples analyzed from the widest ham section.

TABLE 14.3. Days of Sampling after Salting Stage.

Time (Days)			
Batch 1		Batch 2	
BVI	DS I	BI	DS II
1	1	1	1
25	18	25	22
40	49	42	50
206	200	197	194
238	232	222	219

at 3°C and 85% to avoid drying and to allow the porous matrix to absorb the penetrated brine. Afterward, samples A, B, C, and R were taken and analyzed.

After the salting stage, samples were taken at different times throughout the process. Taking into consideration that the initial day ($t = 0$) was just at the end of the salting period, Table 14.3 shows the different times at which samples were taken.

Color measurements were taken at the end of the whole process in the final product. CIEL*a*b* coordinates were measured at the points located in Figure 14.3.

ANALYTICAL CONTROLS

To determine the sodium chloride content, samples were homogenized in a known amount of distilled water at 9000 rpm in an ULTRATURRAX T25 for 5 min and centrifuged to remove any fine debris present in the sample. Afterward, the solution was filtered and exactly a 500-μL aliquot sample was

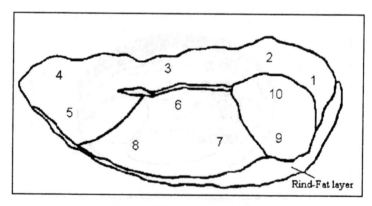

Figure 14.3 Reading points location.

taken and tritated in chloride analyzer equipment (CIBA Corning Mod. 926). Moisture content was determined by oven drying to constant weight at 100°C (ISO R-1442). Color parameters were measured by means of a Minolta spectrophotometer CM-2002. Analytical determinations were conducted in triplicate for each point.

RESULTS AND DISCUSSION

SALTING PROCESS

NaCl content in hams at the end of the salting process, expressed as product weight fraction (x^{NaCl}) and liquid phase mass fraction (z^{NaCl}), are shown in Figures 14.4 and 14.5. The z^{NaCl} value would be more realistic to determine the salt taste and the water activity value (a_w), related with organoleptic and preservation characteristics. The use of brine immersion, with or without vacuum impregnation, implies a faster salting process than the traditional dry salting method. The mean value is higher in both cases, even with a shorter salting period.

The observed NaCl content differences between the two sets of experiments, which are mainly observed in the superficial values, could be due to differences in the salting time, treatment, and in sample thickness, because A, B, and C points are not perfectly defined from ham surface.

The predicted z^{NaCl} values for BVI and BI using Equation (1) for 5.5 and 9 days, respectively, were 0.048 in both cases, whereas the experimental ones obtained were 0.036 and 0.054. BVI results are lower than the expected ones and closer to those that would be predicted for BI, which indicates that the

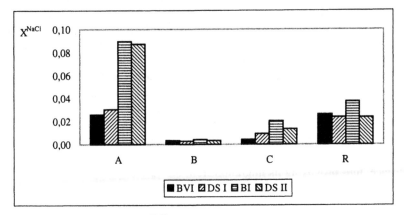

Figure 14.4 The x^{NaCl} values at the end of the salting process.

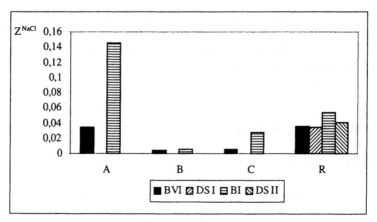

Figure 14.5 The z^{NaCl} values at the end of the salting process.

vacuum pulse length (20 h) was not enough to show significant differences from the BI treatment due to the big sample size. The first practical recommendation would be the use of longer vacuum periods to accelerate the salting process for the atmospheric process.

On the other hand, the z^{NaCl} values obtained for the two batches in the case of DS were lower than that expected for commercial hams (0.042). If the DS experimental results are fitted to Equation (1), 0 being the coordinate b, the slope (k_1 value) would be 0.0039, and the time needed to reach 0.042 g NaCl/g FLP would be 10.8 days. These data would be in accordance with those used in industrial applications (ranging between 1 and 1.5 days per initial kilogram).

The salt content profile analysis (Figure 14.4) shows that the x^{NaCl} content in points A, B, and C for BVI is the lowest for any other salting processes, whereas the NaCl content in the whole ham (point R) is quite big. These observations can lead to the conclusion that during the BVI NaCl penetrates following other pathways than when using the other salting methods, promoted by the HDM action (Fito, 1994).

Weight decrease at the end of the salting process can be observed in Figure 14.6. Similar values are seen for both DS experiments and are higher than the brine salting ones. This could be explained by differences in the chemical potential gradients between ham and the surroundings. In fact, it seems that salt crystals not only need to be dissolved to accomplish the salting process but also have an absorption effect on the liquid phase that appears on the ham surface, thus provoking its migration from the surface. Weight changes at the end of the BVI were lower than for DS I but higher than those for BI. The greater weight decrease in the case of the BVI for the BI in a shorter period of time could be due to initial losses of native ham liquid during the vacuum

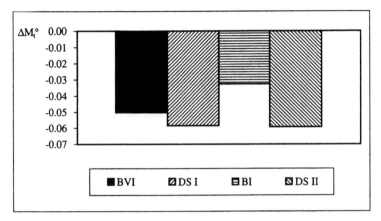

Figure 14.6 Weight changes at the end of the salting process.

period. The liquid phase lost from ham would be mainly composed of blood. Thus, the vacuum pulse would enhance the microbial stability because of blood removal.

POST-SALTING

The post-salting stage is accomplished at refrigeration temperatures to avoid the meat microbial spoilage while NaCl and nitrites and/or nitrates reach the deepest zones. As can be seen in Figures 14.7 and 14.8, during this period, salt penetrates the ham, increasing its content in the B and C points and de-

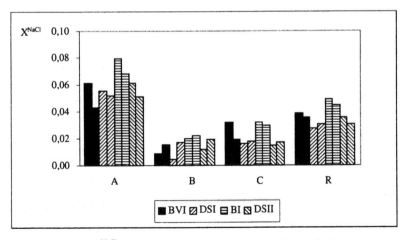

Figure 14.7 The x^{NaCl} values for the two sampling times during the desalting process.

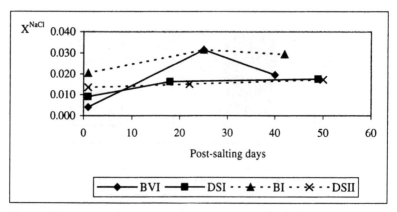

Figure 14.8 NaCl content evolution during the post-salting stage in the deepest point (C).

creasing near the ham surface (point A). This increase in salt content in deeper zones is also favored by water losses during this period in which total sample weight decreases. So, it is observed that the mass transfer mechanisms in the ham interior are quite complex because there are two countercurrent fluxes (mainly water and NaCl, but also protein losses) that lead to a complex a_w profile in the interior of the product.

The post-salting period could be reduced by any method that accelerates the salt content increase in the C point, which would be the most unfavorable one, from the microbial point of view. In Figure 14.8, it can be observed that NaCl content in the C point in BVI and BI reaches its maximum after 22 days of post-salting stage. These values were bigger than that obtained in DS after 50 days, which implies that, in the case of BVI and BI, the post-salting stage could be reduced by at least one half for the traditional method.

DRY-MATURATION

During the dry-maturation period, the global NaCl content increases in the whole ham by concentration due to the drying process. It is also observed that the BI process was too long, because the final NaCl content was nearly 8% (Table 14.4), whereas the average content in commercial hams is approximately 5.5%.

The weight loss that determined the end of the process was 36% (Pierre Poma, 1989), which gives us hams with a moisture content of approximately 50%. These are the experimental values considered in the Spanish cured ham industry.

In Figure 14.9, changes in weight during the processes are plotted against the time for both batches. No significant differences are observed depending on the salting method.

TABLE 14.4. Final Average Values for the Processed Batches.

Treatment	Processing Days	$\Delta M^{\circ}_{finish}$	x^{w}	x^{NaCl}
BV	238	0.3517	0.5183	0.0605
DS I	232	0.3874	0.5456	0.0545
BI	222	0.3267	0.5519	0.0793
DS II	219	0.3452	0.5649	0.0561

COLOR MEASUREMENTS

With the purpose of expressing the visual perception quantitatively, because food color is a factor of decisive quality for its choice and acceptance (Mac-Dougall, 1982; McLaren, 1984; Miller, 1994), color measurements were taken from the cured hams. The results are shown in Figures 14.10 and 14.11.

As is observed in Figure 14.10, there are no differences among the different treatments as to their color evolution. All the hams evolve toward a bigger saturation (red tone), when approaching the central (6 and 10) and deeper areas (8, 7, and 9).

For the three salting treatments, the ham brightness increases when approaching the central (6, 10) and deeper (7, 8, and 9) areas (Figure 14.11). This is due to fat coalition, due to the effect of the thermal treatment, because the most external areas are less rich in these than the internal areas (Buscailhon and Monin, 1994).

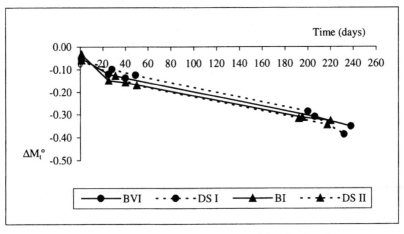

Figure 14.9 Weight decrease during the curing ham elaboration for all treatments.

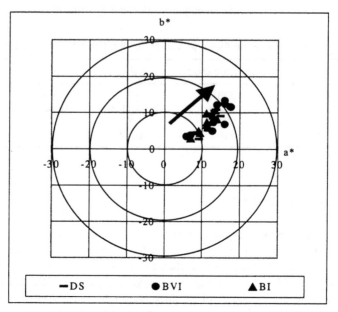

Figure 14.10 Relationship between *a** and *b** for the cured ham.

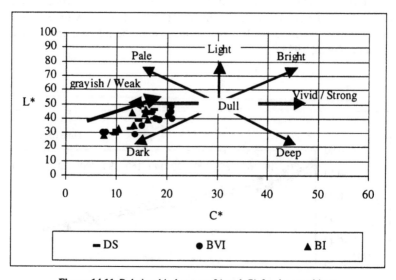

Figure 14.11 Relationship between *L** and *C** for the cured ham.

CONCLUSIONS

Brining of ham implies a reduction in salting time compared with the use of dry salt, especially when working at vacuum pressure. The effect of vacuum pulse for a short time at the beginning of the process seems to be less effective than the continuous vacuum action throughout the brining. Therefore, capillary action promoted at low pressure seems to be the mechanism responsible for the faster salt gain. To obtain the usual salt content reached after industrial salting ($z^{NaCl} \sim 0.042$), the required time in BI, BVI, and DS would be 4.2, 6.9, and 10.8 days according to the linear predictions, which implies a considerable reduction in salting time (of 65 and 48%, respectively).

The post-salting period can be reduced when using the BVI and BI method, because the NaCl content in the deeper points (the riskier ones from the microbial point of view) was higher than those obtained for the DS experiments. The curing process has not suffered significant differences at the end of the process among the three treatments if we observe the results obtained in the studied parameters (NaCl, water, z^{NaCl}, weight losses, and color). This indicates that the brine salting method (at atmospheric pressure or with vacuum impregnation) does not seem to modify the quality of the final product while diminishing the total processing period. The use of brine salting in Spanish cured ham production would contribute to reducing the environmental impact of the process.

NOMENCLATURE

FLP = food liquid phase
M = sample mass (g)
DS = dry salting
BI = brine immersion
BVI = brine vacuum impregnation
x = total mass fraction (g/g)
z = food liquid phase mass fraction (g/g)
t = time (h)
VI = vacuum impregnation
HDM = hydrodynamic mechanism
T = temperature (°C)

Rh = relative humidity (%)
$\Delta M_t^o - \dfrac{M_t^o - M_0^o}{M_0^o}$ = weight change at time t (g/g)

Superscripts:
o = total mass
w = water
NaCl = sodium chloride

Subscripts:
t = values at time t

REFERENCES

Bañon, S., Granados, M. V., Álvarez, D., and Garrido M. D. (1997). *Efecto PSE en el Jamón Curado*. EUROCARNE. Madrid, Spain. 55:27–34.

Barat, J. M., Grau, R., Montero, A., Chiralt, A., and Fito, P. (1998). Feasibility of brining of ham for curing. In: *Meat Consumption and Culture. Congress Proceedings 44th International Congress of Meat Science and Technology.* Ed: Estrategias Alimentarias, S. L. EUROCARNE. Madrid, Spain. pp. 970–971.

Buscailhon, S. and Monin, G. (1994). Factor affecting sensory quality of dry cured ham. II. Effects of raw material quality on quality of dry cured ham. *Viandes et Produits Carnes,* 15(2):39–48.

Chiralt, A. and Fito, P. (1997). Salting of manchego-type cheese by vacuum impregnation. In: *Food Engineering 2000.* Fito, P.. Ortega-Rodríguez, E., Barbosa-Cánovas, G. V. (Eds.). Chapman & Hall, pp. 215–230.

Fito, P. (1994). Modelling of vacuum osmotic dehydration of foods. In: *Water in Foods. Fundamental Aspects and Their Significance in Relation to Processing of Foods.* Fito, P., Mulet, A., Mckenna, B. (Eds.). London: Elsevier Applied Science, pp. 313–328.

Fito, P., Andrés, A., Chiralt, A., and Pardo, P. (1996). Coupling of hydrodynamic mechanism and deformation-relaxation phenomena during vacuum treatments in solids porous food-liquid systems. *J. Food Eng.,* 27:229–240.

Fito, P., Chiralt, A., Serra, J., Mata, M., Pastor, R., Andrés, A., and Shi, X. Q. (1994). An alternating flow procedure to improve liquid exchanges in food products and equipment for carrying out said procedure. Application number 94500071.9. European Patent 0 625 314 A2. Universidad Politécnica de Valencia.

Flores, J. (1996). Mediterranean vs northern European meat products. Processing tecnologies and main differences. *Food Chem.,* 59(4):505–510.

Garcia-Regueiro, J. A. and Diaz, I. (1989). Changes in the composition of neutral lipids in subcutaneous and muscle fat during the elaboration process of spanish cured ham. *Proceedings 35th International Congress of Meat Science and Technology.* Copenhagen, Denmark. August 20–25. Roskilde, Denmark, Danish Meat Research Institute. Vol. III, pp. 719–724.

Gelabert, J., Gou, P., and Arnau, J. (1998). *Disminución del Contenido de sal en el Jamón Curado.* EUROCARNE. Madrid, Spain. 70:27–34.

Iriarte, M. L., Perez, J. A., Gago, M. A., Fito, P., and Aranda, V. (1993). Introducción al estudio del salado de paleta de cerdo por deshidratación osmótica a vacío. In: *Anales de Investigación del Master en Ciencia e Ingeniería de Alimentos.* (Fito, P., Serra, J., Hernandez, E., and Vidal, D., Eds.) SPUPV, Valencia, Spain. Chapter III, pp. 579–593.

MacDougall, D. B. (1982). Changes in the colour and opacity of meat. *Food Chem.,* 9(1/2):75–88.

McLaren, K. (1984). Food colorymetry. In: Walford, J. *Developments in Food Colours - 1.* London: Applied Science Publishers Ltd., Vol. I, Chap. 2, pp. 27–45.

Métodos de ensayo de carnes y productos cárnicos. Determinación de la humedad. Norma UNE 34 552 h2 (ISO R-1442).

Miller, R. K. (1994). Quality characteristics. In: Kinsman, D. M., Kotula, A. W., and Breidenstein, B. *Muscle Foods. Meat, Poultry and Seafoods Technology.* New York: Chapman & Hall. Chap. 11, pp. 296–332.

Ockerman Herbert W., Thomas N. Blumer, and H. Bradford Craig. (1963). Volatile Chemical Compounds in Dry-Cured Hams. Vol. 29, Chap. 2, pp. 123–129.

Pérez Álvarez, J. A., Fernández López, J., Gago Gago, M. A., Pagan Moreno, M. J., Ruiz Pelufo, C., Rosmini, M. R., López Santobeña, F., and Aranda Catalá, V. (1997).

Color properties of dry-cured ham: temperature and pH influence during salting stage. *Journal of Muscle,* 88(3):315–328.

Pierre Pome, J. (1989). La fabricación del jamón curado. Importancia de la congelación de la materia prima. Avances en la tecnología del jamón curado. II *Jornadas Técnicas sobre el Jamón Curado. Instituto de Agroquímica y Tecnología de Alimentos* (IATA). pp. 29–36.

Tapiador Farelo, J. (1989). Influencia de la tecnología de secado en la calidad del jamón curado. II *Jornadas Técnicas sobre el jamón Curado.* pp 47–58.

CHAPTER 15

Salting Studies during Tasajo Making

J. M. BARAT
G. ANDUJAR
A. ANDRÉS
A. ARGÜELLES
P. FITO

INTRODUCTION

TASAJO is a traditional Cuban intermediate moisture product made from beef sheets that are soaked in brine, drained, rubbed in salt, and layered with salt in vats for several days, washed, and sun-dried (Andújar and Valladares, 1989). The obtained product has a characteristic composition with average values of $x^w = 0.3$, $x^{NaCl} = 0.23$, $x^p = 0.4$, $x^{fat} = 0.05$. Tasajo is stored under normal atmospheric conditions (high temperature and relative humidity) and finally is desalted and minced before its consumption.

During the process, fermentation and oxidation processes occur (Andújar and Valladares, 1989; Pinto et al., 1998), which in addition with the high salt content and reduced moisture, contribute to the characteristics of the final product. The production method is strongly artisan, and the introduction of new technologies could be interesting to optimize the process.

Salting is one of the main processes undergone by meat. During this step, the product gains NaCl and loses water and soluble solids (mainly soluble proteins). The understanding of the process and the use of new technologies are essential to improve this basic step in the processing chain.

BASIC KNOWLEDGE

It is well known that the addition of sodium chloride to meat changes the protein isoelectric point and the water-holding capacity (Wilding et al., 1986). The exact mechanism by which meat water-holding capacity increases is not fully understood. It has been suggested that the bonded Cl^- ions in the proteins provokes its repulsion, which implies the meat swelling after the A band has been dissolved because of the NaCl action, thus provoking the external solution uptake by capillary forces (Offer and Trinick, 1983). Nevertheless, a big increase in the medium ionic force can provoke protein denaturalization, even reaching the point of its precipitation. Because the sodium chloride content reaches very high values during Tasajo salting, a complete protein denaturation at the end of the process can be expected.

In general, water losses imply volume shrinkage, thus compressing the structure. When driving forces related with differences in concentration between meat and its surroundings disappear, the structure stress tends to be released and brine is gained. The fact that there is an increase in weight at the end of the process during brine salting and not during dry salting gives consistency to this idea (Andújar et al., 2000). This behavior has been observed in the case of fruit during osmotic dehydration (Barat et al., 1998).

In this work, the salting step during Tasajo making is analyzed. The traditional salting method (dry salting: DS) has been studied and compared with brine salting working at atmospheric pressure (brine immersion: BI) and with the initial application of a vacuum pulse (brine vacuum impregnation: BVI).

MATERIALS AND METHODS

Fresh beef semitendinosus muscle was used in two sets of experiments, and the brine salting process was accomplished under stirring conditions. In the first set, whole muscles (w.m) were used working at 6°C. The aim of this experiment was to characterize the usual dry salting method and compare with the two brine salting methods proposed (BI and BVI). Weight and compositional changes were determined in triplicate in all cases. The experiments finished after approximately 120 h, thus being the moment that corresponded to the maximum weight loss.

In the second set of experiments, muscle pieces (m.p) (8 × 8 × 2.5 cm) were used. The influence of vacuum period during vacuum impregnation (VI) (1 and 2 h), temperature (2, 6, and 10°C) and sample thickness (1, 2, and 2.5 cm) on weight changes was studied for brine and dry salting.

ANALYTICAL DETERMINATIONS

For the sodium chloride determination, samples were homogenized in distilled water at 10,000 rpm in an Ultraturrax T25 (Janke & Kunkel, Staufen, Germany) for 7 min and centrifuged to remove any fine debris present in the sample. An aliquot of centrifuged sample was taken and titrated in chloride analyser equipment (Sherwood Mod. 926, Cambridge, UK). Moisture content was quantified by oven drying for 1 h at 125°C (AOAC, 1980).

RESULTS AND DISCUSSION

MASS BALANCES

Weight changes during meat salting can be attributed to three main fluxes: water outflow, meat-soluble solids outflow (mainly proteins), and NaCl gain. The complete mass balance can be observed in Equation (1), and the calculated weight changes were determined as indicated in Equations (2) and (3). Meat-soluble solids losses were determined by means of Equation (1) using the experimental total water and NaCl weight changes. The obtained results, depending on process time, can be observed in Figure 15.1. It can be seen that meat-soluble solid losses can be neglected as compared to water losses and NaCl gain.

$$\Delta M^o_t = \Delta M^w_t + \Delta M^{NaCl}_t + \Delta M^{ss}_t, \tag{1}$$

$$\Delta M^o_t = (M^o_t - M^o_0)/M^o_0 \tag{2}$$

$$\Delta M^i_t = (M^o_t \cdot x^i_t - M^o_0 \cdot x^i_0)/M^o_0 \tag{3}$$

Figure 15.2 shows the close fitting of the global balance Equation (1) to all the experimental results obtained when meat-soluble solid losses are neglected. This indicates that the experimental determinations were right and that for further explanations the meat protein losses will not be taken into account.

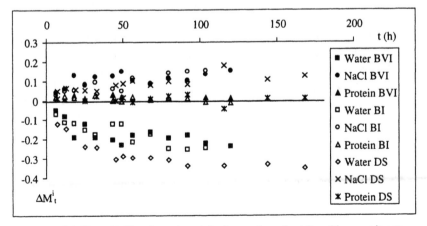

Figure 15.1 Water, NaCl, and protein weight changes throughout the salting experiments.

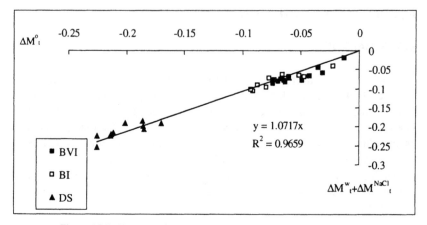

Figure 15.2 Checking of mass balances for all salting experimental data.

WEIGHT CHANGES DURING TASAJO SALTING

The traditional salting operation (using dry salt) was characterized by using whole sheets (as usual) and meat pieces ($8 \times 8 \times 2.5$ cm). The use of brine salting, with or without vacuum impregnation, instead of dry salting was studied.

Total weight changes during the salting experiments can be seen in Figure 15.3. Important differences, depending on the salting method, can be observed. The vacuum-impregnated samples are those that have a bigger yield. Initially, they increase their weight because of the brine uptake due to the vacuum pulse. After the initial weight increase, it decreases because of the more intense dehydration for NaCl uptake. The differences in the behavior of BI and BVI treatments can be attributed to the observed initial weight increase due to the vacuum pulse.

The DS treatment implied a higher weight decrease for the wet salting methods. This can be explained by the higher sample dehydration. During DS, an initial water outflow would appear to dissolve the salt crystals of the surface. Once a thin liquid phase layer appears, the Na^+ and Cl^- ions penetrate the meat, but the dehydration process continues while the liquid phase moves toward the bed of salt crystals because of capillary forces. This phenomenon, in addition to the dilution effect in the thin layer close to the meat surface during brine salting, contributes to the more intense dehydration phenomenon in DS for the brine salting methods and thus to a more marked weight decrease and a lower process yield.

Changes in process yield due to the different salting methods are quite evident. In the case of a salted and dried product, part of the retained liquid

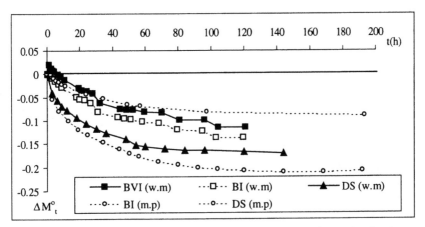

Figure 15.3 Weight changes throughout the salting experiments for whole meat (w.m) and meat pieces (m.p).

phase can be released by dripping during the drying process, hence affecting the observed differences in the process yields.

Another important phenomenon observed during the salting experiment was that the whole meat (w.m) diminished their weight in a different way than the meat pieces (m.p). Differences in the main fiber orientation in relation to mass transfer fluxes and the changes in the specific surface could explain the different behavior. Changes in the meat fiber orientation imply that, in the case of whole meat, it is parallel to the main flux direction and that, in the case of meat pieces, there is a higher proportion of fibers perpendicular to the lateral meat planes. The two above-mentioned changes would imply a lower weight decrease during BS than to the whole meat because of the increase in NaCl uptake for water loss, and in the case of DS, the effect woud be the contrary.

Another important difference between dry and brine salting is that the standard deviation during BI and BVI is much bigger than that observed for DS (Figure 15.4). On the other hand, there is a common observation among the three salting methods; the standard deviation increases with time.

The standard deviation gives an idea of the different behavior for the three meat pieces used in each experiment. The above-mentioned differences are only dependent on differences in the meat samples (composition, size, weight, and structure) because the other experimental conditions remain the same for all of them. Another important question is that the role of the aforementioned differences in meat in the behavior depends on the magnitude of the driving forces compared with the common driving forces generated between the meat piece

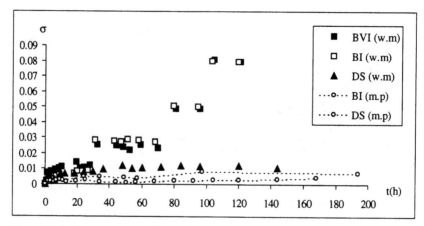

Figure 15.4 Changes in the standard deviation (σ) during salting experiments for whole meat (w.m) and meat pieces (m.p).

and its surroundings. The increase in the standard deviation with time can be explained if it is considered that the driving forces between meat and its surroundings due to the presence of salt crystals or saturated brine decrease when the salting process advances and the equilibrium situation is being reached. The decrease in those driving forces makes it possible to perceive the other ones due to differences in sample size, composition, and/or structure.

The greater standard deviation value for brine salting would be related to a lower initial tendency of water to leave meat compared with the dry salting process, as is reflected by the final weight decrease. It is also important that in the case of brine salting, the pressure supported by meat is lower than in the case of dry salting. This lower pressure and the presence of a liquid phase surrounding the meat piece would enable the meat piece partially to recover its volume, thus increasing its weight at the end of the experiment as was observed.

By comparing the standard deviation for whole meat and meat pieces, important differences are observed between them. For meat pieces, the size, shape, and weight are more similar, and thus the contribution of those factors to standard deviation is minimized, being lower than for the whole meat. Although the standard deviation is lower, it is still higher for brine salting than for dry salting.

VACUUM PERIOD DURATION INFLUENCE ON SALTING PROCESS

The VI process can provoke the brine uptake because of the hydrodynamic mechanism (Fito, 1994) or to mechanical effects on the structure due to the changing system pressures. The total amount of external solution gained by the sample is strongly dependent on some process parameters: temperature, solution viscosity, and pressure value. Among these, one important variable is the period of time that the sample is submitted to under atmospheric pressure before the atmospheric pressure is restored. The vacuum period is necessary to eliminate the gas phase and thus create spaces into which the solution would be pumped when restoring the initial pressure.

Weight changes throughout Tasajo salting vacuum periods of 1 and 2 h are shown in Figure 15.5. It can be observed that the longer vacuum period implies a lower weight decrease, in accordance with the expected results. A higher NaCl penetration is accomplished with a longer vacuum pulse and so a less dehydrated product is obtained. Another remarkable observation is that, although the weight change pathway is different, the two curves are nearly parallel, and so it would imply that the subsequent process kinetics does not change.

TEMPERATURE INFLUENCE ON THE SALTING PROCESS

Salting experiments were conducted for three of the possible process temperatures (2, 6, and 10°C), and results can be seen in Figure 15.6. Although differences in temperatures were not very important, the salting pathway was strongly affected. The higher and lower weight losses were obtained working at 10 and 6°C, respectively. There are two possible effects of temperature; on one hand, it affects mass transfer coefficients and on the other the protein matrix, thus affecting the water-holding capacity. Further studies would be necessary to fully understand the temperature effect.

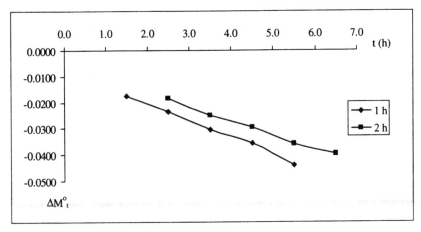

Figure 15.5 Weight changes throughout Tasajo salting at different vacuum period times (1 and 2 hours).

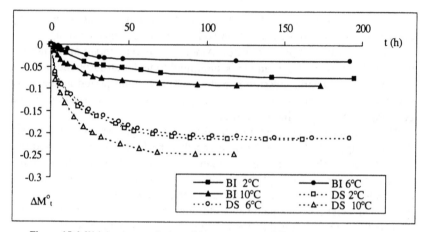

Figure 15.6 Weight changes during salting treatment at different process temperatures.

From a process point of view, 10°C might be interesting when working with DS because of the higher temperature (lower process cost) and higher dehydration effect (shorter subsequent air-drying process). Both effects would be beneficial from an economic point of view. Nevertheless, microbiological and yield studies are needed to determine if 10°C can be considered the optimum process temperature.

SAMPLE THICKNESS INFLUENCE ON SALTING BEHAVIOR

Because Tasajo is desalted and minced before its consumption, it means that one of the process variables that can be easily modified if interesting is sample thickness. The influence of this variable on weight changes was studied by working with meat pieces. The results obtained in weight changes throughout the process for different sample thicknesses can be seen in Figure 15.7. It is observed that the salting process is strongly affected by this variable. The observed differences affect kinetics and the maximum weight decrease. In DS, the rate of weight changes is faster when lower thickness samples are used, and the maximum weight decrease is lower. This would imply faster processes and higher yields after the salting process. Both effects are beneficial from the economic point of view. Even more, the lower sample thickness could lead to a faster drying and desalting process.

In the case of the BI, the sample behavior is quite different from the DS. Samples of 2.5- and 2-cm thickness behave in a similar way, but the maximum weight decrease in the 2-cm samples is lower. One-cm thickness sample behavior is very different from the other ones. The salting step in this case is faster (the maximum weight loss is achieved earlier), and the maximum weight loss is more significant than in the case of the thicker samples. These

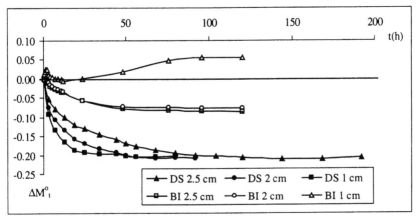

Figure 15.7 Effect of thickness during BI and DS working at 6°C with meat pieces (8 × 8 cm).

different pathways could be explained by the more important effect of superficial phenomena, such as the increase in water-holding capacity due to the initial NaCl uptake, which would explain the initial weight gain. The subsequent weight decrease would be explained by the same mechanisms as in the other cases; greater water losses than the NaCl uptake. The final weight increase, at the end of the process, would be explained by the volume recovery near the equilibrium by sample swelling. The release of the mechanical accumulated or generated stress during sample shrinkage due to water losses implies the uptake of the external solution. This behavior at the end of the process would be similar to that observed for fruits (Barat et al., 1998).

It can be observed that the weight loss for meat pieces of different thicknesses was different from those when working with whole meat, even when sample thickness was changed. This would imply that process yield would be higher when working with whole meat in DS and with meat pieces in brine salting. Nevertheless, if a highly dehydrated product is desired, the best choice would be to work with small meat pieces and DS, because of the faster process and the more intense weight loss, which implies a more important water loss and hence the following subsequent air-drying process would be shortened.

SALTING KINETICS

Salting kinetics could be related with two main phenomena: weight changes and liquid phase NaCl concentration. Weight changes are mainly related with the process yield, and liquid phase concentration with preservation (a_w) and organoleptic aspects.

Weight changes occurring until the maximum weight loss is reached can be easily modeled by using Equation (4). The k_1 parameter is related with the process kinetics, whereas the k_2 parameter is related with the displacement of the fitted equation from the ideal diffusion behavior. The salting process, as can be observed in Table 15.1, affects both parameters.

$$\Delta M_t^o = k_1 \cdot \frac{t^{0.5}}{l} + k_2 \tag{4}$$

Weight decrease rate (k_1 parameter) depends on two mass fluxes: water outflow and NaCl uptake. It is much faster when working with DS because of the faster water loss and bigger [water loss/NaCl gain] ratio compared with the brine-salting methods. When comparing BI and BVI, the k_1 parameter is quite similar. The faster weight decrease when working with BI can be explained if it is taken into account that the driving force is reduced in the vacuum impregnated samples. This reduction would be due to the higher initial brine uptake by hydrodynamic mechanism (Fito and Pastor, 1994).

Differences in the k_2 parameters can also be explained by the salting method used. In the case of DS, a fast initial water loss is needed to dissolve the solid salt surrounding the meat. This fast initial dehydration is the reason why the fitted line has negative k_2 value. In the case of BI and BVI, the k_2 value is greater than 0 in both cases. These values are explained by the initial brine uptake because of the HDM action and the changes in the water-holding capacity in the superficial layer due to the initial NaCl uptake and the subsequent brine gain. The k_2 value is greater in the case of BVI because of the initial weight increase by VI.

Changes in the FLP concentration can be modeled by considering that the main transfer mechanism is that of diffusion. A simplified form (considering only the first term of the equation) of the integrated equation of the Fick's law for plane sheets and short-time processes [Equation (5)] was used to obtain the apparent diffusion coefficients (Crank, 1975). The K term in Equation (5) was introduced to consider other mass transfer mechanisms than that of diffusion.

$$1 - Y_t^{NaCl} = \left(\frac{z_t^{NaCl} - z_o^{NaCl}}{z_e^{NaCl} - z_o^{NaCl}} \right) = 2 \cdot \left(\frac{D_e \cdot t}{\pi \cdot l^2} \right)^{\frac{1}{2}} + K \tag{5}$$

The apparent diffusion comparing BI and BVI for whole meat, and DS and BI for meat pieces was calculated by plotting the ($1 - Y^{NaCl}_t$) values vs. $t^{0.5}/l$

TABLE 15.1. Kinetic Parameters of
Weight Changes ($T = 6°C$).

	DS	BI	BVI
k_1 (mm/h$^{0.5}$)	−0.259	−0.223	−0.216
$k_2(-)$	−0.012	0.012	0.033
r^2	0.99	0.98	0.98

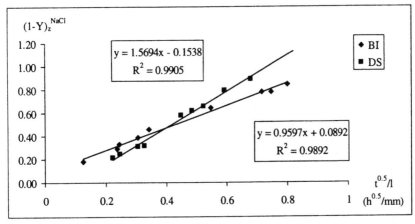

Figure 15.8 $(1 - Y^{NaCl}_t)$ values versus $t^{0.5}/l$ for salting of meat pieces salting ($T = 6°C$).

(Figure 15.8). It must be taken into account that the diffusion coefficient determined in this way is affected not only by the mass transfer kinetics but also by the (water outflow:NaCl uptake) ratio. It is necessary to keep this in mind to analyze the obtained results.

The calculated FLP diffusion coefficients are shown in Table 15.2. It can be seen that the diffusion coefficient is higher for BVI than for BI, because of the use of the vacuum pulse at the beginning of the process, thus favoring the NaCl uptake. On the other hand, the diffusion coefficient for DS is higher than for BI probably because of the more intense water loss. On the other hand, when comparing the diffusion coefficient for BI working with whole meat and meat pieces, an increase in that value for meat pieces can be observed because of the higher specific surface of the samples.

The K values are in accordance with the phenomena explained in the previous points. An initial dehydration in DS without an important NaCl uptake implied a negative K value, whereas the positive K value for BI and BVI would

TABLE 15.2. Diffusion Coefficients ($T = 6°C$).

	DS	BI	BVI
		Whole Meat	
D_e (m²/s)		$1.39 \cdot 10^{-10}$	$1.74 \cdot 10^{-10}$
$K(-)$		0.0722	0.1267
		Meat Pieces	
D_e (m²/s)	$5.37 \cdot 10^{-10}$	$2.01 \cdot 10^{-10}$	
$K(-)$	-0.1538	0.0892	

be due to the brine uptake caused by the increase in water-holding capacity at the beginning of the process.

CONCLUSION

The salting step in Tasajo making has been studied. Three different salting procedures (dry salting, brine immersion, and brine vacuum impregnation) were compared. Two main mass fluxes were observed in all cases: the water outflow and the NaCl uptake. The protein losses can be neglected compared with the other mass fluxes. The more intense dehydration phenomena takes place during the dry salting process because of the different salting medium (salt crystals) compared with the other two salting methods (saturated brine), thus implying a lower process yield. Variability was always higher in the case of the wet salting methods, probably because of the more important effect of the meat structure during processing. Process variables assayed were sample thickness, vacuum period duration and temperature; these variables showed an important influence on meat weight changes throughout the salting exper-iments. Salting kinetics was well modeled with the proposed model, and the kinetic parameters were obtained and analyzed. Further studies are necessary to determine the influence of the different salting methods on the subsequent steps before product consumption, drying, storage, and desalting.

NOMENCLATURE

a_w = water activity
BI = brine immersion
BVI = brine vacuum impregnation
D_e = effective diffusivity (m^2/s)
DS = dry salting
HDM = hydrodynamic mechanism
l = half thickness of dehydrated product (m)
M = sample mass (kg)
m.p = meat pieces
T = temperature (°C)
t = time (s)
VI = vacuum impregnation
w.m = whole meat
x^j = mass fraction of j in food (kg. j/total kg)

Y = reduced driving force
z^j = mass fraction of j in food liquid phase
σ = standard deviation

Superscripts:
fat = fat
NaCl = sodium chloride
o = total
p = protein
ss = soluble solids
w = water

Subscripts:
0 = initial values ($t = 0$)
t = values at time t

REFERENCES

Andujar, G. and Valladares, C. (1989). Study of a traditional intermediate moisture meat product: "Tasajo" I. Processing method and chemical composition. In *Proceedings. 35th International Congress of Meat Science and Technology.* Copenhagen, Denmark. Roskilde, Denmark: Danish Meat Research Institute.

Andújar, G., Argüelles, A., Barat, J. M., Andrés, A., and Fito, P. (2000). Tasajo salting by brine immersion. Influence of vacuum impregnation. In *Engineering and Food at ICEF 8.* Technomic Publishing Co., Inc., Lancaster, PA (in press).

AOAC. 1980. *Official Methods of Analysis of the AOAC.* AOAC, Washington. DC.

Barat, J. M., Chiralt, A., and Fito, P. (1998). Equilibrium in cellular food osmotic solution systems as related to structure. *J. Food Sci,* 63(5):836–840.

Crank, J. (1975). *The Mathematics of Diffusion.* Oxford University Press, Oxford.

Fito, P. (1994). Modelling of osmotic dehydration of foods. In *Water in Foods. Fundamental Aspects and Their Significance in Relation to Processing of Foods.* Fito, P., Mulet, A., Mckenna, B., eds. Elsevier Applied Science, London, pp. 313–328.

Fito, P. and Pastor, R. (1994). On some non-diffusional mechanisms occuring during vacuum osmotic dehydration. *J. Food Eng.,* 21:513–519.

Offer, G. and Trinick, J. (1983). On the mechanism of water holding in meat: the swelling and shrinking of myofibrils. *Meat Science,* 8:245–281.

Pinto, M. F., Ponsano, E. H. G., Campos, S. D. S., Franco, B. D. G. M., and Shimokomaki, M. (1998). Charqui meats are fermented meat products. In *Meat Consumption and Culture,* Vol. II. Driesde, A. and Monfort, J. M., Eds. EUROCARNE. C.20:858–859.

Wilding, P., Hedges, N., and Lillford, P. (1986). Salt-induced swelling of meat: The effect of storage time, pH, ion-type and concentration. *Meat Science,* 18:55–75.

CHAPTER 16

Application of Vacuum Impregnation Technology to Salting and Desalting Cod (*Gadus morhua*)

A. ANDRÉS
S. RODRÍGUEZ-BARONA
J.M. BARAT
P. FITO

INTRODUCTION

SALTING fish is both a method of preserving and a preliminary operation to some smoking, drying, and marinating processes. Bachalao, salted cod (*Gadus morhua*), is a popular traditional food in Mediterranean cuisine, and about 10% of the world production is salted after its capture. Traditionally, it is imported in salted form and then desalted to cook traditional dishes.

Meat and fish salting is one of the oldest treatments in food preservation that consists of the diffusion of salt throughout the muscular tissues. The traditional production of salted cod is a simple process, taking approximately 2–8 weeks, depending on the type of curing desired in the finished products. The salted cod has been produced for centuries, during which time the production protocols have remained remarkably unchanged.

Dried fish quality is affected by many factors such as the quality of fish before salting, ratio of salt to fish, the ambient temperature, the salting method, and the method and degree of drying after salting (Ismail and Wootton, 1992). There are three common methods for fish salting (FAO, 1981):

(1) Kench salting: Dry salt crystals are rubbed into the flesh after which the fish is staked. While the salt penetrates, the extracted moisture drains away. This method is suitable for lean fish such as cod.
(2) Pickling: Dry salt crystals are rubbed into the flesh but the moisture extracted while the salt is penetrating is not drained away, thus the fish is subsequently immersed in a salt pickle of extracted fluids. This method is more common for fat species (Boeri et al., 1982; Nesvadba, P., 1999).
(3) Brining: This is a wet salting method in which the fish is soaked in a con-

185

centrated salt solution. The salt content of the final product can be regulated by controlling the salting time and brine temperature.

Salted cod reaches a salt concentration of nearly 20% (w/w), and the water content decreases from 80% to about 65%. The end product shows a high microbiological stability and specific organoleptic features.

Although these methods are the ones most commonly used, some innovative techniques to speed up salting can be found in the literature (Ismail and Wootton, 1992). Salt penetration into the fish flesh ends when the salt concentration in the aqueous phase of the tissue becomes equal to that in the surrounding solution. At higher salt concentrations, fish muscle loses water because of salt denaturation of the muscle proteins. Thus, there is a critical salt concentration (8–10% w.b.) above which the protein is denatured rapidly and salt gain is accompanied by water loss. Before this critical concentration, an initial uptake of both salt and water takes place. Temperatures between 0 and 20°C had no effect on the critical salt level, although this level is reached faster at high temperature (Ismail and Wootton, 1992).

Salted cod, dried to a greater or lesser degree, plays an important role in some Mediterranean countries such as Spain, Portugal, and Greece, and it is used in the preparation of many traditional dishes. Salted cod is mainly imported from Norway and Iceland where it is a product of great economic and cultural interest.

The desalting process, which is a necessary step before either the drying operation or the definitive cooking, is largely traditional, and no scientific knowledge exists about it. Usually, the salted cod is soaked in tap water for at least 24 h, and the desalting process is conducted at room temperature. During the last few years, an important decrease in salted cod consumption has been appreciated principally because of a different life style. Consumer trends are leaning toward eating easy-to-prepare or ready-to-use products, and for this reason a more profound knowledge of the desalting process is necessary to commercialize a ready-to-use product, with homogeneous composition and an adequate quality.

In both the salting and desalting processes, diffusion is the major mass transfer mechanism responsible for sodium chloride transport, and it is well known that diffusional mechanisms are quite slow. It was shown that the application of vacuum impregnation (VI) accelerated the mass transfer phenomenon in porous solid-liquid systems as is the case of cod/brine and cod/water in salting and desalting processes, respectively. The application of VI results in a permanent replacement of intercellular gases by liquid media as a consequence of the hydrodynamic mechanism. VI has appeared recently as a new technology (Fito, P., 1994; Fito and Pastor, 1994) suitable for improving mass transfer kinetics in solid-liquid systems. A short vacuum pulse at the beginning of the process has been applied successfully in cod and trout salting (Del Saz et al., 1994; Escriche et al., 1996). After applying vacuum for some minutes, the hydrodynamic mechanism acts while the atmospheric pressure is restored. Occluded gas in the product pores is expelled during the vacuum period, and when pressure increases again, pores are quickly filled with the surrounding solution, increasing the weight and mass transfer surface with the liquid phase (Fito and Pastor, 1994). So, VI applied to the brining salting process may increase the mass transfer kinetics. The influence of vacuum treatment on the kinetic of the mass transfer phenomena is very important, especially where concerning water loss and weight reduction (Chiralt and Fito, 1997; Fito and Chiralt, 1997). On the other hand, because cod desalting also implies working with a solid-liquid system, and wet cod is a porous product, the use of VI also seems to be interesting to improve process yield and mass transfer kinetics. These two processes imply important mass transfer phenomena, thus the knowledge of the process kinetics is very important for an adequate industrial application. The influence of VI on both cod salting and desalting operations has been analyzed by comparing the results with those obtained by traditional methods.

VACUUM SALTING

Mass changes during the salting process are the result of two opposite fluxes: water loss and salt uptake. Weight variations of the samples during the salting process can be expressed as the relative mass variation ΔM_t calculated as:

$$\Delta M_t = \frac{M_t - M_0}{M_0} \tag{1}$$

where

M_0 = initial mass sample (g)
M_t = mass sample after a time (t) of treatment (g)

Figure 16.1 shows the relative mass change obtained from cod fillets salted by different procedures (brining salting with and without vacuum pulse (50 mbar) and kench salting with and without pressing the fillets ($P = 183$ kg/m^2)). The duration of the vacuum pulse (0, 30, 60, and 120 min) and the presence or not of the central bone in the fillet are variables that affect both the kinetic and the final result of the process.

It can be observed that mass equilibrium has been reached in all cases, and its value depends on the salting method according to the final liquid fraction content. Salt concentration of the liquid fraction at the equilibrium is assumed to be the same as that corresponding to a saturated brine.

It is interesting to observe that all the ΔM_t curves present a common shape similar to those obtained in osmotic dehydration of fruits. This typical behavior can be described in two steps. During the first one, ΔM_t decreases dra-

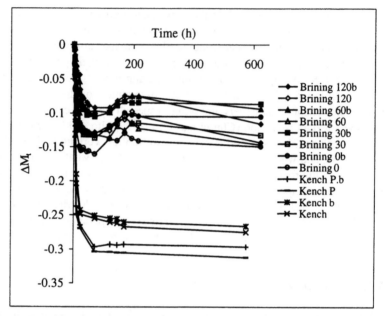

Figure 16.1 Relative mass variation throughout time, 0, 30, 60, and 120 refers to the duration of the vacuum pulse (min); b refers to samples salted with their bone and, P indicates kench salting with pressure ($P = 183$ kg/m^2).

matically because of a fast initial water loss. This fact, along with a collapse of the muscle fibers because of the osmotic stress, leads the system to a pseudo-equilibrium state in which samples are of minimum weight. This situation is followed by the fiber relaxation allowing the inflow of a certain amount of osmotic solution until the true equilibrium is reached (Barat et al., 1998).

Differences between fillets with and without the central bone are quite important in brine salting and very small in kench salting. This can be attributed to a filling up of the central bone by the salting solution in the case of brine salting and also to the contribution of the bone weight in ΔM_t calculations. The influence of pressing or not pressing the fillets during kench salting is on the yield of the process. The draining away of moisture in this case is more efficient as a consequence of the pressure effect, leading to a lower yield of the desalting process.

Comparing the two salting methods, remarkable differences in mass loss can be found between kench and brine salting. In this work, the most relevant difference between the two salting methods will be the overall yield of the salting process, but from reports by other authors, brine salting offers other kinds of advantages such as the antioxidant role of the brine (Boeri et al., 1982). Yield differences are in accordance with the different liquid fraction-solid matrix ratio (LF/SM) at the equilibrium state (Rodriguez-Barona et al., 2001a).

VACUUM DESALTING

Desalting process is affected by many factors such as water changes, quality of water, temperature, pressure, and how the cod was salted, postmortem time before salting of cod, etc. Another factor that could be expected to affect the kinetic of the desalting process would be the water stirring that is performed to avoid the film effect at the cod surface. Results from experiments under different stirring conditions and their influence on mass transfer revealed that the water stirring rate does not bring about any significant differences on cod mass changes (Rodriguez-Barona et al., 2001b). This is an important conclusion that could simplify the design of industrial devices for fish desalting, where moving water can be used but not necessarily to reduce the desalting time.

As in other solid-liquid processes, the results of the cod-desalting process would be affected by pressure gradients in the system. Vacuum desalting of cod could be conducted by applying a short vacuum pulse to impregnate the fish structure and continuing the process at atmospheric pressure. Figure 16.2 shows the mean values of total mass variation after an hour of desalting for samples tried with different times of vacuum pulse.

Some experiments varying the duration of the vacuum pulse and the process temperature have been conducted to analyze the influence of these two variables throughout the first hour of the process. Significant differences in mass increment can be observed between samples submitted to a vacuum pulse of 15 and 30 min, those desalted without vacuum pulse and the ones submitted to a 5-min vacuum pulse. This indicates that 15 min is the equilibrium time for sample impregnation. The same progression, with fewer differences, is observed for volume changes. It is important to point out that the optimum duration of the vacuum pulse is affected by the sample size and the bigger the piece, the longer the time needed to impregnate the sample completely. Cod samples used for these trials were in the shape of small cubes ($2 \times 1 \times 1$ cm^3). The weight increase promoted by vacuum impregnation is quicker during the first hour of the desalting and changes occur more slowly near the equilibrium time. Nevertheless, the yield of the process is always bigger when vacuum pulse is applied. The influence of the vacuum pulse on texture would be more appreciated and, in this sense, more studies are needed.

Another critical aspect of cod desalting is the differences between the characteristics of the salted cod fish as raw material. Salting procedures differ not only from country to country but also from factory to factory, leading to different characteristics of this product. Some comparative experiments performed on salted cod from Norway and from Spain showed that these differences greatly affect the kinetic and the yield of the desalting process and the final quality of the desalted cod (Figure 16.3).

The process yield was higher for the Spanish cod and practically independent of the desalting method. The vacuum desalting method was more efficient in the case of Norwegian cod and affected both the kinetic and the process

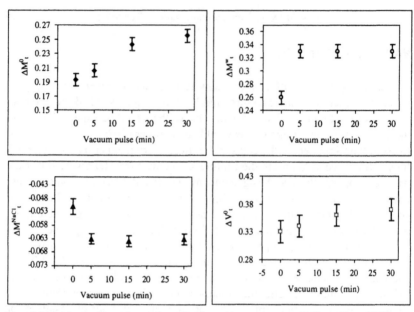

Figure 16.2 Mean values of mass, water, salt, and volume changes (ΔM^0_t, ΔM^w_t, ΔM^{NaCl}_t, and ΔV^0_t) after 1 h of desalting for different vacuum pulse durations.

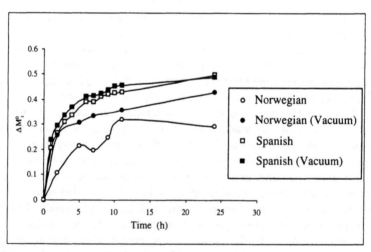

Figure 16.3 Relative mass changes during desalting process of Norwegian and Spanish salted cod with and without vacuum pulse.

yield. These differences would be connected to the different structure of the protein depending on the salting procedure.

The preliminary results show that vacuum technology would be more interesting from the point of view of the process yield than from the kinetic one. Vacuum desalting would also affect final texture and the microbial quality, but it needs more studies.

ACKNOWLEDGEMENTS

The authors acknowledge the European Union (FAIR program, PL 98-4179) for the financial support given to this investigation.

REFERENCES

Barat, J. M., Chiralt, A., and Fito, P. (1998). Equilibrium in cellular food osmotic solution systems as related to structure. *J. Food Sci.* 63(5):836–840.

Boeri, R. L., Moschair, S. M., and Lupín, H. M. (1982). Salting of hake (*Merluccius hubbsi*). A comparative study of pickle and kench salting. *Rev. Agroquím. Tecnol. Aliment.* 22(1):139–145.

Chiralt, A. and Fito, P. (1997). Salting of Manchego type cheese by vacuum impregnation. In *Food Engineering 2000*, P. Fito, E. Ortega-Rodriguez, and G. V. Barbosa-Cánovas, eds. London: Chapman & Hall, pp. 215–230.

Del Saz, A., Mateu, A., Fito, P., and Serra, J. A. (1994). Vacuum salt impregnation of cultivated trouts (*Salmo gairdneri*). In *Proceedings of the Poster Sesion. International Symposium of the Properties of Water, Practicum II.* Argaiz, López-Malo, Palou, and Corte, eds. Puebla, Mejico: UDLA, pp. 57–60.

Escriche, I., Serra, J. A., Fito, P., and Rivero, E. (1996). Estudio de la influencia de la deshidratacion osmotica a vacio en el salado del bacalao (*Gadus morhua*). In *Anales del I Congreso Ibero-Americano de Ingenieria de Alimentos.* A. R. Matos Chamorro, P. J. Sobral, A. Chiralt, and S. M. Alzamora, eds. Valencia, España: SPUPV, pp. 142–151.

FAO (1981). The preservation of losses in cured fish. *FAO Fish. Rep. 279.* Rome.

Fito, P. (1994). Modelling of osmotic dehydration of foods. In *Water in Foods. Fundamental Aspects and Their Significance in Relation to Processing of Foods.* P. Fito, A. Mulet, and B. Mckenna, eds. Londres: Elsevier Applied Science, pp. 313–328.

Fito, P. and Chiralt, A. (1997). Osmotic dehydration: An approach to the modeling of solid-liquid food operations. In *Food Engineering 2000*, P. Fito, E. Ortega-Rodriguez and G. V. Barbosa-Cánovas, eds. London: Chapman & Hall, pp. 231–252.

Fito, P. and Pastor, R. (1994). On some non-diffusional mechanisms occurring during vacuum osmotic dehydration. *J. Food Eng.* 21:513–519.

Ismail, N. and Wootton, M. (1992). Fish salting and drying: A review. *ASEAN Food Journal.* 7(4):175–182.

Nesvadba, P. (1999). Osmotic treatments of fish and fish products. In *Osmotic Treatments for the Food Industry,* A. M. Sereno, ed. Porto: FEUP ediçoes, pp. 59–66.

Rodriguez-Barona, S., Barat, J. M., Andrés, A., and Fito, P. (2001a). Cod salting by vacuum impregnation. In *Engineering and Food for the 21st Century*, J. Welti-Chanes, G. V. Barbosa-Cánovas, and J. M. Aguilera, eds. Lancaster, PA: Technomic Publishing Co., Inc. (in press).

Rodriguez-Barona, S., Ibáñez, J. B., Andrés, A., Barat, J. M., and Fito, P. (2001b). Cod desalting by vacuum impregnation. In *Engineering and Food for the 21st Century*, J. Welti-Chanes, G. V. Barbosa-Cánovas, and J. M. Aguilera, eds. Lancaster, PA: Technomic Publishing Co., Inc. (in press).

PART V

VACUUM IMPREGNATION, OSMOTIC TREATMENTS AND MICROWAVE COMBINED PROCESSES

CHAPTER 17

Use of Vacuum Impregnation in Smoked Salmon Manufacturing

G. BUGUEÑO
I. ESCRICHE
A. CHIRALT
M. PÉREZ-JUAN
J. A. SERRA
M. M. CAMACHO

INTRODUCTION

IN the salted/smoked salmon-manufacturing process by traditional methods, such as salting with dry salt or brining, the required time is very long, ranging between 10–30 days for dry salting (Tanikawa et al., 1985) and between 24–48 h for brining (Jarvis, 1987). Important microbiological alterations can occur when salting is prolonged over several days. After salting, the surface skin is dried, and the smoking process can be performed by different procedures. Tanikawa et al. (1985) report a smoking time length of 2 weeks at increasing temperatures between 18 and 25°C, whereas Jarvis (1987) points to different smoking times ranging between several hours and days at 20–26°C, depending on the type of product. Throughout smoking, microbiological problems can be increased, mainly in lengthy process (Himelbloom et al., 1998).

The movement of salt in fish muscle can be properly described as a diffusion process (Nesvadba, 1997), but scarce data have been published on diffusion of salt in fish. Peter (1971) reports a diffusivity value in fish muscle for NaCl of 2.10^{-9} m^2/s at room temperature. Diffusion mechanisms are very slow and accelerating the salting process injection of salt in fish fillets has been proposed. Nesvadba (1990) models the diffusion of brine injected into fish filets. Injection takes advantage of hydrodynamic mechanisms to improve brining efficiency.

Recently, another hydrodynamic mass transfer mechanism (HDM), which takes place during solid-liquid vacuum operations, has been described and modeled (Fito, 1994; Fito et al., 1993; Fito and Pastor, 1994). This hydrodynamic mechanism occurs when a porous product is immersed in a liquid phase

195

and the temperature or pressure changes. This mechanism is controlled by the presence of occluded gas inside the porous structure of the product, which is compressed or expanded according to pressure or temperature changes that allow the external liquid to enter or exit the pores (Fito and Pastor, 1994).

BVI consists of two steps: first a vacuum pressure was held in the brine tank during a time t_1 and afterward the atmospheric pressure was restored and salmon remains a time t_2 in the brine. During the vacuum period, the gas occluded in the porous structure of the product was expanded and partially flows out, allowing a more intense capillary penetration. In the second period, the restoring of atmospheric pressure promotes the residual gas compression and the external brine penetration, thus allowing a faster salting process (González et al., 1999).

The aim of this work is to study the salt penetration kinetics by conventional brine immersion (BI) and by brine vacuum impregnation (BVI) in salmon fillets, as well as the comparison of some physicochemical parameters of salted salmon obtained by both procedures.

EXPERIMENTAL

MATERIALS

Salmon (*Salmo salar*) fillets (4 × 5 cm, 1-cm thick), skinned and boned, were obtained from salmon purchased in a local market. Brine for both treatments, BI and BVI, was prepared at 370 g/L containing 1.5% liquid smoke. Each sample was prepared in triplicate.

BRINE VACUUM IMPREGNATION AND BRINE IMMERSION TREATMENTS

BI and BVI processes were conducted for different times (ranging from 0 to 300 min) to analyze kinetics. In BVI process, a pressure of 50 mbar was applied for the first 5 min in the tank, afterward restoring the atmospheric pressure. Experiments were performed at 5, 15, and 25°C. At each process time, three fillets were removed from solution, and their surfaces were gently blotted with tissue paper. Initial and final sample weight, moisture, and salt content were determined in each sample.

Kinetics of brining was evaluated through the development of weight loss (ΔM), water loss (ΔM_w), and salt gain (ΔM_s), calculated according to Equations (1)–(3), in process time.

$$\Delta M = \frac{m_t - m_o}{m_o} \qquad (1)$$

$$\Delta M_w = \frac{m_t x_{wt} - m_o x_{wo}}{m_o} \qquad (2)$$

$$\Delta M_s = \frac{m_t x_{st} - m_o x_{so}}{m_o} \qquad (3)$$

THE EFFECTIVE POROSITY

Salmon effective porosity to the action of hydrodynamic mechanism was evaluated according to a previously reported gravimetric methodology (Fito and Pastor, 1994). Isotonic solution of NaCl (a_w = 0.990, 17.4 g NaCl/L) was used as impregnating solution. The samples (2 $\times$ 2 $\times$ 2 cm salmon cubes) were immersed in this solution for 15 min at vacuum pressure (50, 100, 250, and 400 mbar), followed by 15 min at atmospheric pressure. Initial (m_o) and final (m_f) sample weight was recorded, the difference being the mass of impregnated solution. Taking into account the solution density (ρ_s) and the sample volume (from its dimensions), the sample volume fraction occupied by the isotonic solution (X) was estimated as a function of the vacuum pressure applied.

ANALYTICAL DETERMINATIONS

For the sodium chloride determination, samples were homogenized in distilled water at 9000 rpm in an Ultraturrax T25 (Janke & Kunkel, Staufen, Germany) for 5 min and centrifuged to remove any fine debris present in the samples. An aliquot of centrifuged sample was taken and treated in chloride analyzer equipment (Sherwood Mod. 926, Cambridge, UK). Moisture content (x_w) was quantified by vacuum drying the samples at 60°C until constant weight was achieved (AOAC 20013 standard method).

RESULTS AND DISCUSSION

DEVELOPMENT OF A_W DURING SALTING PROCESS: COMPARISON BETWEEN EXPERIMENTAL AND PREDICTED VALUES

Figure 17.1 shows the development of a_w values of the different salted samples as a function of NaCl molality determined in their liquid phase (simplified to water plus salt) from salt content and moisture data. Predicted values of a_w applying Pitzer equation (Pitzer, 1973; Pitzer and Moyarga, 1973) were also plotted for each molality. It can be observed that experimental a_w values are slightly lower than the predicted ones. So, predicted values can be used as a safe criteria for product stability. To correct small deviations, compari-

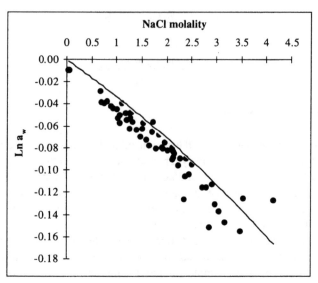

Figure 17.1 Development of sample a_w as a function of NaCl molality in the product liquid phase (water plus salt). Line: predicted values by Pitzer equation.

son of experimental and predicted a_w values is conducted by means of a linear regression, which was highly statistically significant. The Equation (4) obtained can be used to obtain a more accurate value of the sample a_w value than those directly predicted by Pitzer equation.

$$a_{wexp} = 1.034 + 0.0439 a_{wPitzer} \quad (r^2 = 0.948) \tag{4}$$

VACUUM IMPREGNATION BEHAVIOR OF SALMON TISSUE

Figure 17.2 shows the experimental impregnation data plotted in a linear way according to impregnation model (Fito, 1994) as a function of the vacuum pressure applied [Equation (5)]. A value of X of 5.5% was obtained when 50 mbar was applied at the vacuum step. According to Equation (5), straight line might have an intercept equal to 0. Nevertheless, sample viscoelastic behavior and the coupled sample deformation due to pressure changes can provoke deviations from Equation (5) (Fito et al., 1992). From the X values obtained, salmon brining by vacuum impregnation can be expected to be more effective than the process conducted at atmospheric pressure.

$$X = \epsilon_e \left(1 - \frac{1}{r}\right) \tag{5}$$

$$(r \approx p_{atm.}/p_{vac.}) \tag{6}$$

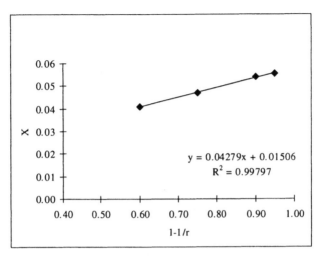

Figure 17.2 Sample volume fraction impregnated by an isotonic external liquid as a function of the vacuum pressure applied in the first vacuum step.

SALTING KINETICS OF SALMON FILLETS

Two aspects are relevant in the kinetic model of salting. The rate of composition change of the product liquid phase, which defines the salting time required to ensure a determined product stability or a_w, and the rate of product weight loss, defined by the relative fluxes of water and salt, which determines the final process yield or product weight loss. For kinetic analysis of liquid phase composition changes in salmon, this is considered as a binary system composed by water and salt. The reduced driving force for brining at each process time has been defined by Equation (7) in the mass fraction of water or salt in the liquid phase (z) and will have the same value in composition of either component. In Equation (7) the equilibrium value of z has been considered equal to the mass fraction of the respective (water: w or salt: s) component in brine (y) (e.g., $z_{se} = y_s$).

$$Y_w = Y_s = \frac{z_{st} - z_{se}}{z_{so} - z_{se}} = \frac{z_{wt} - z_{we}}{z_{wo} - z_{we}} \qquad (7)$$

Considering salmon fillet geometry as an infinite plane sheet, a simplified Fickian equation with only one term of the series solution for short times [Equation (8)] can be used to estimate the effective diffusion coefficient (D_e) from the experimental composition data. Figure 17.3 shows the plot $1 - Y_s$ vs. $\sqrt{t}$ for brining experiments conducted at different temperatures at atmospheric pressure (BI) and by applying vacuum for 5 min at the beginning of the process (BVI). The straight lines fitted to the points in Figure 17.3 showed a significant value of the intercept (k_0 in Table 17.1). This may be attributed

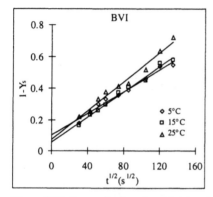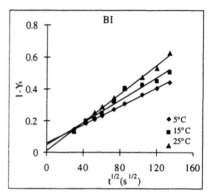

Figure 17.3 Changes in the reduced brining driving force for BVI and BI processes of salmon fillets at different temperatures.

to a fast composition change occurring in the sample at short times ($t \rightarrow 0$), due to impregnation by the vacuum pulse action, or by capillary effects in processes conducted at atmospheric pressure. From the slope, D_e values were obtained, taking into account the sample thickness (Table 17.1).

$$1 - Y_s = \frac{z_{st} - z_{so}}{y_s - z_{so}} = 2\sqrt{\frac{D_e t}{\pi l^2}} \qquad (8)$$

As can be seen in Table 17.1, the D_e values were greater for BVI than for BI processes at a determined temperature. Figure 17.4 shows the Arrhenius plot for D_e values in both treatments, which reflects a greater temperature influence on the kinetic parameter when the process was conducted at atmospheric pressure. Activation energy (E_a) values were 31 and 19 kJ/mol, respectively, for BI and BVI. Likewise, the k_0 values were slightly higher for BVI than for BI treatments. The faster salt uptake, less temperature dependent, in the salmon liquid phase in BVI may be explained in the contribution of HDM to salt transport. The lower temperature influence on the D_e values suggests that these values involve not only the transport due to diffusional

TABLE 17.1. Kinetic Constant k_0 and Effective Diffusion Coefficient (D_e) for BVI and BI of Salmon.

Treatment	k_0	$D_e \times 10^{10}$ (m²/s)	r^2
BVI—5°C	0.101	2.36	0.980
BVI—15°C	0.056	3.15	0.988
BVI—25°C	0.074	4.15	0.979
BI—5°C	0.062	1.61	0.999
BI—15°C	0.047	2.45	0.963
BI—25°C	0.012	3.93	0.995

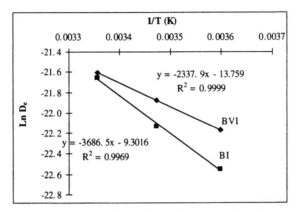

Figure 17.4 Arrhenius plot of diffusivity values in BI and BVI treatments.

mechanisms but also that induced by hydrodynamic fluxes on which temperature exerts only a mild influence. From the kinetic data, a prolonged action of HDM throughout the whole brining time seems to be inferred.

This could be explained in a progressive sample volume relaxation, after the compression induced when atmospheric pressure is restored. Sample volume recovery would imply a suction effect on the external solution due to generated internal pressure gradients (Barat et al., 1999) and so hydrodynamic flux coupled with the salt diffusional uptake.

Salt gain (ΔM_s) and water (ΔM_w) and weight (ΔM) losses, all referred to the sample initial mass [Equations (1)–(3)] has been empirically modeled by Equations (9)–(11) in the square root of time ($\sqrt{t}$). Figure 17.5 shows an ex-

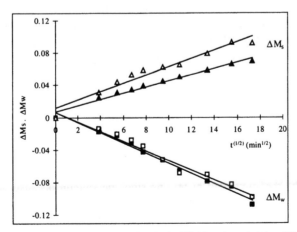

Figure 17.5 Kinetics of water loss and salt gain for BI (closed symbols) and BVI (open symbols) processes at 5°C.

ample of the close fitting of Equations (9) and (10) to the experimental data at 5°C. Table 17.2 shows the values of kinetic constants for each case together with the respective r^2 values.

$$\Delta M_w = K_w\sqrt{t} + K_{wo} \tag{9}$$

$$\Delta M_s = K_s\sqrt{t} + K_{so} \tag{10}$$

$$\Delta M = K_o + K\sqrt{t} \tag{11}$$

From the fitted model, it is possible to estimate process time and the corresponding process yield and final sample composition by fixing the final product water activity that ensures its stability. Figure 17.6 shows a flowchart for this estimation process. The parameter values obtained for a determined a_w that is the usual value in commercial salmon ($a_w = 0.965$) are included. This value corresponds to a salt concentration in the product liquid phase $z_s = 0.019$, as deduced by Equation (4), that was reached at 5°C after 80 and 50 min, respectively, for BI and BVI processes. At this concentration, the weight loss obtained in a BVI process was 3%, whereas 2% was obtained in BI at the same temperature. This value was associated with a greater moisture content (63%) in the product obtained by BVI than in that obtained by BI (62%), which will imply a juicier product.

For other quality parameters of salted salmon, such as color and texture, no significant differences between color coordinates of newly salted BI and BVI products have been found. Nevertheless, mechanical properties, related with product texture, show small differences between the two products. BVI samples appeared firmer (with greater deformability modulus), showing a greater resistance to the plastic flow that occurs at greater force and relative deformation than in BI samples (Bugueño, 1999). These differences could be related with the deeper salt penetration in the tissue in BVI, thus affecting the aggregation state of myofibrils in the muscle tissue or to the vacuum pulse effect that promotes the expulsion of free internal liquid and gas phases, giving rise to a more compact structure. Nevertheless, no sensory perception of these differences was uncovered (Bugueño, 1999).

CONCLUSION

BVI can be recommended in brining-smoking salmon because process is shortened while product retains more water and weight at a determined level of its water activity reduction. Likewise, the product does not seem to suffer any losses in its quality attributes.

TABLE 17.2. Kinetic Constants for Salt Gain and Water and Weight Losses in Different Salmon Brining Treatments.

Treatment	$K_v \times 10^2$ (min$^{-0.5}$)	$K_{wo} \times 10^2$	r^2	$K_s \times 10^2$ (min$^{-0.5}$)	$K_{so} \times 10^2$	r^2	$K \times 10^2$ (min$^{-0.5}$)	$K_o \times 10^2$	r^2
BVI—5°C	-0.59	0.64	0.975	0.52	1.14	0.954	-0.37	0.40	0.977
BVI—15°C	-0.89	0.18	0.989	0.53	0.77	0.978	-0.52	0.07	0.983
BVI—25°C	-1.20	-0.38	0.974	0.58	1.11	0.963	-0.59	0.23	0.977
BI—5°C	-0.62	0.56	0.985	0.38	0.74	0.975	-0.46	1.00	0.953
BI—15°C	-0.96	2.34	0.996	0.41	1.27	0.925	-0.62	0.51	0.989
BI—25°C	-1.41	2.93	0.996	0.49	1.05	0.978	-0.74	0.57	0.987

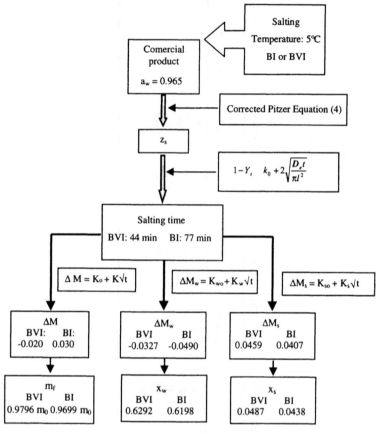

Figure 17.6 Flowchart for the application of kinetic model to estimate process time and yield and product composition in BI and BVI process to give product water activity to the desired level.

NOMENCLATURE

D_e = effective diffusion coefficient (m²/s)
a_w = water activity
ΔM = weight loss (kg/kg)
ΔM_w = water loss (kg/kg)
ΔM_s = salt gain (kg/kg)
r = compression ratio (p_2/p_1)
t = time (s)
Y = reduced driving force
m_o = initial weight sample (kg)
m_t = sample mass at time t (kg)

x_{jo} = mass fraction of j in initial sample (kg j/kg sample)
x_{jt} = mass fraction of j in the sample at time t (kg j/kg sample)
x_{je} = mass fraction of j in the sample at equilibrium (kg j/kg sample)
z_{jt} = mass fraction of j in the sample liquid phase at time t ($t = 0,....$)
z_{je} = mass fraction in the sample liquid phase of j at equilibrium
y_j = mass fraction of j in brine
X = volume fraction impregnated of the initial sample (m^3/m^3)
ϵ_e = effective porosity
ρ_s = Density of external solution (kg/m^3)

ACKNOWLEDGEMENTS

Authors thank to Comisión Interministerial de Ciencia y Tecnología (Spain), FAIR Program (E.U, DGXII), and CYTED program the financial support.

REFERENCES

Barat, J. M., Albors, A., Chiralt, A., and Fito, P. 1999. Equilibration of apple tissue in osmotic dehydration: Microstructural changes. *Drying Technology,* 17(7&8): 1375–1386.

Bugueño, G. 1999. Salado ahumado de salmón (*Salmo salar*) por impregnación a vacío. Influencia del envasado en la calidad. Ph.D. thesis, Universidad Politécnica de Valencia.

Fito, P. 1994. Modelling of vacuum osmotic dehydration of food. *Journal of Food Engineering,* 22:313–328.

Fito, P. and Pastor, R. 1994. Non diffusional mechanisms occurring during vacuum osmotic dehydration. *Journal of Food Engineering,* 21:513–519.

Fito, P., Andrés, A., Pastor, R., and Chiralt, A. 1993. Vacuum osmotic dehydration of fruits. In *Process Optimisation and Minimal Processing of Foods.* Singh, P., Oliveira, F. Eds. CRC Press, Boca Ratón, pp 107–121.

Fito, P., Shi, X. Q., Chiralt, A., Acosta, E., and Andres, A. 1992. Vacuum osmotic dehydration of fruits. In *ISOPOW-V.* Valencia. Spain.

Gonzàlez, C., Fuentes, C., Andrés, A., and Fito, P. 1999. Effectiveness of vacuum impregnation brining of manchego type curd. *International Dairy Journal,* 9:143–148.

Himelbloom, B., Crapo, C., and Pfutzenreuter R. 1998. Microbial quality of an Alaska native smoked salmon process. *Journal of Food Protection,* 59(1):56–58.

Jarvis, N. R. 1987. *Curing of Fishery Products.* Teaparty Books, Kingston, MA.

Nesvadba, P. 1990. Diffusion of salt injected into fish. *Torry Document* 2316.

Nesvadba, P. 1997. Osmotic treatments of fish and fish products. *Osmotic Treatments for the Food Industry.* 1st O.T. Seminar, Porto, Portugal. Ed. Alberto M. Sereno. Concerted Action Fair-CT-96-1118.

Peter, G. R. 1971. Diffusion in a medium containing a solvent and solutes with particular reference to fish muscle. PhD thesis, University of Aberdeen.

Pitzer, K. S. 1973. Thermodynamics of electrolytes I. Theoretical basis and general equations. *Journal of Physical Chemistry,* 77:268–277.

Pitzer, K. S. and Moyarga, G. 1973. Thermodynamics of electrolytes II. Activity and osmotic coefficients for strong electrolytes with one or both ions univalent. *Journal of Physical Chemistry,* 77:2300–2308.

Tanikawa, E., Motihoro, T., and Akiba, M. 1985. Marine products in Jaoan, Revised Edition. Koseisha Koseikaku Co., Ltd., Tokyo.

CHAPTER 18

Combined Osmotic and Microwave-Vacuum Dehydration of Apples and Strawberries

U. ERLE
H. SCHUBERT

INTRODUCTION

OSMOTIC dehydration is a gentle way of removing water from plant tissues such as fruits or vegetables. More than 50% of the water is taken out with the help of concentrated sugar solutions. After that, the fruit pieces are very soft and still subject to spoilage by a variety of microorganisms. The water content needs to be lowered further to gain microbiological stability without cool storage.

In this study, microwave-vacuum dehydration has been used to achieve shelf stability. The products obtained had high-quality taste and color. Retention of vitamin C—a substance of high nutritional value—was determined as a quality indicator. Volume after osmotic treatment alone and after combined treatment was measured to show how the osmotic treatment progresses and how the properties of the final product can be influenced. Monitoring the changes in sugar composition during the osmotic process gives interesting information on the complicated nature of this unit operation that comprises more than just mass transfer phenomena.

THEORY

In osmotic dehydration, pieces of fruit or vegetable are immersed in an aqueous solution. Sucrose or (cheaper) mixtures of sugars are normally used for fruits. Because the cell membranes only allow very limited transfer of sugars into the tissue, equalizing the concentrations of dissolved substances inside and outside the fruit takes place by movement of the water from the inside to the outside. The material may also lose a portion of its own solutes (vitamins, volatiles, minerals, etc.). Figure 18.1 gives an overview of these transport phenomena.

Thermal damage is much reduced in microwave-vacuum drying, because most of the time, temperatures are only slightly above the boiling point at the selected pressure. At 50 mbar, the pressure applied in this study, the boiling point of pure water is 32.9°C. In the final stages of microwave drying, the temperature may reach 80°C, but thermal damage in this period is still relatively low, because heat sensitivity decreases with decreasing water content. With vacuum drying, heat transfer is usually a problem and, therefore, a limitation to attainable drying rates. Microwaves overcome this "bottleneck," because the heat does not need to be conducted. Instead, it is generated in the tissue. This allows for energy transfer rates much higher than in conventional drying operations, especially in the falling rate period (Roussy and Pearce, 1995). The water within a sample is mainly responsible for the absorption of microwave power. As a consequence, the distribution of power in microwave drying is self controlled, at least to a certain extent. Those areas that received more power than others dry quicker and, therefore, will absorb less power from then on. However, when the power level is set too high, overheating and burning of dry areas, a so-called "thermal runaway," may occur. It is caused by the dielectric behavior of biopolymers such as cellulose. The dielectric loss of dry cellulose increases with temperature as soon as a critical temperature is exceeded.

The applicaton of an osmotic treatment before microwave-vacuum drying combines some advantages of both unit operations. Because no phase transition takes place in osmotic dehydration, energy consumption is especially low, even if the diluted solution needs to be reconcentrated by evaporation. Microwaves require electricity, a relatively expensive form of energy (in most

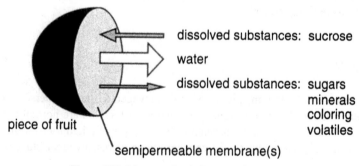

Figure 18.1 Mass transfer during osmotic treatment.

countries), but they are only used in the final stages of drying, where they can be used more efficiently than hot air (Gunasekaran, 1999).

When untreated pieces of apple or strawberry are microwave-vacuum dried, they will normally shrink, sometimes as much as air-dried samples. Osmotic pretreatments provide a tool for incorporating sucrose in the food, which may already contain pectins. These are known to form gels with sucrose (Pilnik, 1980). The resulting improvements in volume are the main subject of this study. Another beneficial effect is that the addition of sugars in general means that less water needs to be removed for shelf stability.

EXPERIMENTAL

OSMOTIC TREATMENT OF APPLES

Apples (*Golden Delicious*) were cored, peeled, and cut in twelve slices, which were treated osmotically. Approximately 400 g of apple slices (12–18 cm^3) were put into a stirred 60% (w/w) sucrose solution (see Figure 18.2). It was tested in earlier experiments that the quality of mixing in this setup ensures that the influence of mass transfer outside the fruit is negligible. The ratio of apples to solution was 1:9. Osmotic treatment was conducted for durations ranging from 2 to 25 hs. After treatment, samples were dipped into water five times, blotted in kitchen paper, weighed, and used for measurements or microwave-vacuum drying.

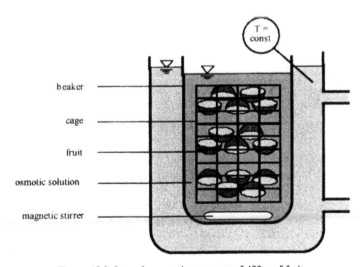

Figure 18.2 Setup for osmotic treatment of 400 g of fruit.

OSMOTIC TREATMENT OF STRAWBERRIES (FOR SUGAR AND VITAMIN C MEASUREMENTS)

Measurements of sugar concentration and vitamin C tend to be quite scattered when they are conducted with different batches or with the same batch of fruit with different degrees of ripeness. Hence, a bigger vessel for osmotic dehydration has been used in these experiments. It is capable of holding more than 20 samples of 100 g from one batch resulting in much reduced scattering of the data compared with earlier measurements, in which a new batch of fruit had to be used for each duration of osmotic treatment. One objective here was to detect possible enzymic activity due to osmotic treatment in a sucrose solution, which would cause changes in the sugar composition of the strawberries. In addition, the sucrose uptake against time had to be documented.

Twenty-two samples of 100 ± 1 g of medium sized strawberry halves (4–8 cm^3) were put in nets and dried in a stirred 60% sucrose solution (20 kg) at 20°C. After treatment, samples were dipped into water three times, blotted in kitchen paper, weighed, and frozen for later analysis or microwave-vacuum dried for vitamin C measurements of the final product. Three more samples were just dipped in a 60% sucrose solution and then treated like the other samples to serve as reference.

OSMOTIC TREATMENT OF STRAWBERRIES (FOR VOLUME MEASUREMENTS)

The small vessel (see Figure 18.2) was used for the osmotic treatment of strawberries in those experiments concerning shrinkage. Four hundred g of medium sized strawberry halves were treated in 60% sucrose solution for 3–22 h at 20°C.

MICROWAVE-VACUUM DRYING

Microwave drying of the fresh and osmotically treated fruits took place in a pilot-scale microwave plant, designed to simulate an industrial microwave tunnel (see Figure 18.3). The pieces were put on a tray that allows the steam to exit in all directions. The trolley goes back and forth, and the magnetrons are only switched on when it comes near the corresponding horn antenna, so the microwaves are used in a pulsed way. Because microwave power absorption depends mainly on the amount of water in the samples, the microwave dryer was always fed with samples containing 140 ± 5 g of water, regardless of the percentage of water removed in the osmotic process. This makes the experiments comparable. However, one of the results in earlier experiments was that the osmotic pretreatment made microwave drying a little more ef-

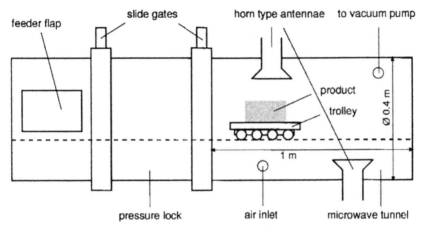

Figure 18.3 Microwave pilot plant.

fective both in removed water and achieved water activity. The main reasons for this are higher dielectric losses caused by the sucrose (Padua, 1993) and better microwave absorption of less shrunken objects. As a result, the following microwave programs were applied: strawberries: 390 W for 37 min plus 195 W for 15 min; apples: 390 W for 30 min plus 195 W for 39 min.

Untreated, fresh apple and strawberry pieces needed longer microwave drying. Water activities lower than 0.6 were achieved in all cases. Pressure was always 50 mbar.

MEASUREMENTS OF SUGAR AND VITAMIN C

Frozen samples were homogenized in a laboratory mixer together with 100 g of ice and 50 ml of 15% meta-phosphoric acid. The slurry was poured into a flask, and its pH was altered to 3.5–4 by potassium hydroxide. The flask was then filled with water up to 500 ml. The contents of the flask were filtered and used for the following measurements.

Fructose, glucose, and sucrose were determined by HPLC. Vitamin C in strawberries was measured by an enzymic test (Boehringer no. 409 677). Because the vitamin content in the apples used here proved to be rather low, a different method had to be chosen: vitamin C (ascorbic acid + dehydroascorbic acid) in fresh, osmotically treated and microwaved apples was measured by a chemical method involving titration with dichlorphenolindophenol-sodium. A 1% solution of oxalic acid was used in the homogenizer to inhibit further degradation of the vitamin C. After homogenization, the slurry was filtered, and the chemical procedure was applied to the filtrate.

VOLUME MEASUREMENTS

Volume of the fresh and osmotically treated strawberries and apples was determined by measuring their buoyancy in water, whereas the volume of dry samples was found through measuring the displacement of glass spheres of known bulk density.

RESULTS AND DISCUSSION

The gain of sucrose during osmotic dehydration of strawberries is depicted in Figure 18.4. The first triple point at 0 h osmotic treatment—the samples were just dipped in the sucrose solution and then washed immediately—gave a starting point of 5% sucrose based on the initial dry mass. Most of the following sucrose gain of approximately 20–25% (initial dry basis) happened within in the first 2 h.

The evolution of the fructose and glucose content is presented in Figure 18.5. There was no significant increase of the two sugars in the first few hours. After approximately 7 h, an upward trend for both glucose and fructose started. This was most likely caused by hydrolysis of sucrose. Because the sucrose content remained more or less constant in that period of time, the conclusion is that the rate of sucrose conversion into glucose and fructose was the same as the rate of sucrose entering the tissue. The flow of sucrose seemed to have stopped (see Figure 18.4), but in reality it had not. Models describing osmotic dehydration should take into account that enzymic reac-

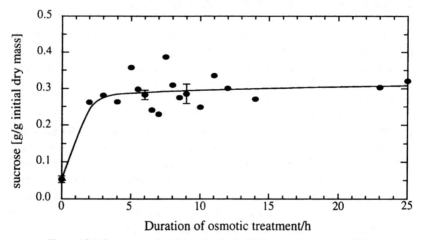

Figure 18.4 Sucrose uptake of strawberries in 60% sucrose solution at 20°C.

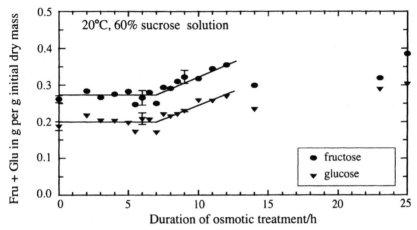

Figure 18.5 Evolution of fructose and glucose during osmotic dehydration of strawberries in 60% sucrose solution at 20°C.

tions may—after a few hours—influence the sugar composition and, therefore, the whole process.

Preservation of vitamin C is the subject of Figure 18.6. During the osmotic treatment (upper curve), a loss of a few percent—comparable with typical losses during storage at 20°C—occurred for long durations. After microwave-vacuum drying with the procedure described earlier, approximately 60% of the vitamin was still detected, regardless of the duration of osmotic treatment.

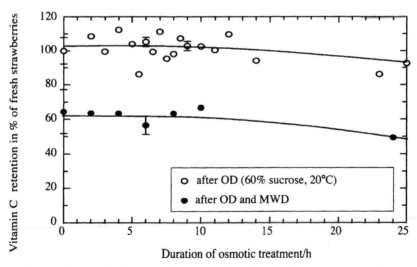

Figure 18.6 Vitamin C retention during osmotic and combined drying of strawberries.

Single experiments with more gentle procedures (not reported here) indicate that preservation of the vitamin can still be improved, but only at the price of longer drying times.

Figure 18.7 shows that the apples exhibited virtually no loss of vitamin C during osmotic treatment. The final vitamin C content was approximately 60% with no obvious correlation to the duration of the osmotic treatment. As with the strawberries, this value is only valid for the microwave procedure applied here.

The concentration of vitamin C in these apples was quite low even in the fresh material (25 mg/100 g of initial dry matter), which is why the enzymic test had to be replaced by a chemical method. In this particular case, the retention of the vitamin is not very important for nutritional reasons, but it still serves as an indicator of thermal damage.

In Figure 18.8, the upper curve simply represents shrinkage caused by water loss during osmotic dehydration of strawberries. The lower curve shows how much an osmotic pretreatment can influence the volume of the final product. Strawberries that stayed in the sucrose solution for 22 h took up a volume higher by a factor of 2.5 compared with those solely microwave dried. It is apparent that this increase cannot be explained by the volume of the extra sucrose itself. Samples without pretreatment were also notably softer than those with a long osmotic step. The latter seemed more crispy and brittle. It is very likely to be the interaction of the sucrose with the pectins in the strawberries that has caused this change in structure and volume.

Apple slices were also treated osmotically under the same conditions as strawberries. The result, given in Figure 18.9, was very similar to that of the

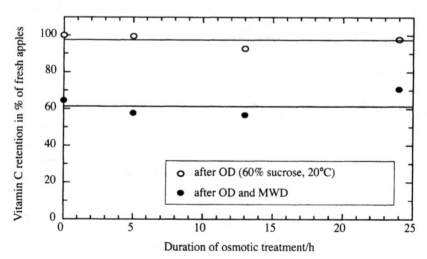

Figure 18.7 Vitamin C retention during osmotic and combined drying of apples.

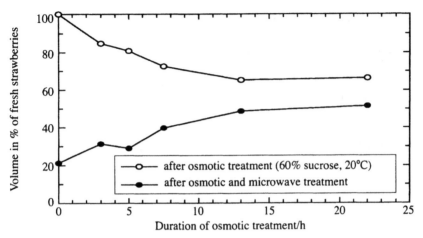

Figure 18.8 Shrinkage during osmotic and combined drying of strawberries at 20°C.

strawberries. Again, the volume of the pieces after microwave drying could be increased by a factor of up to 2.5 (lower curve).

CONCLUSIONS

Microwave-vacuum drying of osmotically pretreated fruits combines the benefits of both unit operations. Selecting the conditions during osmotic treat-

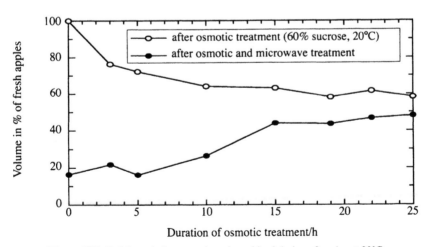

Figure 18.9 Shrinkage during osmotic and combined drying of apples at 20°C.

ment offers the possibility of influencing both the efficiency of microwave dehydration and the properties of the final product. High-quality products in color, taste, vitamin C content, and low shrinkage can be obtained. Vitamin C retention was approximately 60% for apples and strawberries with the microwave program used in this study. In the combined process, volume was preserved by up to 50% of the fresh apples and strawberries.

ACKNOWLEDGEMENT

The authors thank the Max-Buchner-Forschungsstiftung for the support of this research.

REFERENCES

Gunasekaran, S. 1999. Pulsed microwave-vacuum drying of food materials. *Drying Technology* 17(3):395–412.

Padua, G. W. 1993. Microwave-heating of agar gels containing sucrose. *J. Food Sci.* 58(6):1426–1428.

Pilnik, W. 1980. Pektine und alginate. In: *Gelier- und Verdickungsmittel in Lebensmitteln (Gelling and Thickening Agents in Foods)*. Forster Verlag AG, Zürich/Switzerland.

Roussy, G. and Pearce, J. A. 1995. *Foundations and Industrial Applications of Microwaves and Radio Frequency Fields*. John Wiley & Sons, Chichester/Great Britain.

CHAPTER 19

Application of Microwave Drying after Osmotic Dehydration: Effect of Dielectric Properties on Heating Characteristics

E. TORRINGA
F. LOURENCO
I. SCHEEWE
P. BARTELS

INTRODUCTION

OSMOTIC dehydration can be used as an effective method to remove water from vegetable tissues while simultaneously introducing solutes in the product, Figure 19.1. With this dehydration technique, shelf stability cannot be obtained. Moisture removal by evaporation at intermediate moisture content after osmotic dehydration should be applied to reach a lower final moisture content for achievement of shelf stability.

Microwave energy has successfully been utilized in different unit processes in the food industry, such as pasteurizing of convenience food and thawing (Torringa, 1996). Compared with conventional processes, microwaves can improve the product quality because of the increased heating rate. An increase of nutritional content, an improvement of taste, or a more favorable texture may be obtained. Microwaves penetrate into the material and cause a volumetric heating, resulting in a rapid heating of the product (Nijhuis et al., 1998). For moist materials, the heat transfer is practically not hindered by conduction from the surface to the center of the product. Furthermore, this rapid heating may result in an internal pressure buildup producing a very porous material (Lepourhiet and Bories, 1980).

General advantages for applying microwaves in vegetable dehydration are:

- reduction of drying time/increase of efficiency of the drying process
- reduction of thermal and moisture gradients
- decrease of overdrying of outer layers/more homogenous heating
- preservation of porosity/improved rehydratation

217

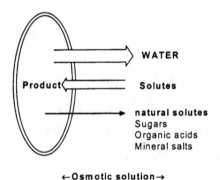

Figure 19.1 Principles of osmotic dehydration.

- reduction of shrinkage
- decrease of energy consumption

Reasons for applying osmotic treatment before microwave dehydration are:

- solute uptake brings about more homogenous heating by microwaves due to modified dielectric properties; for instance, less overheating of the central parts of concave shaped mushrooms in microwave drying
- possibility to produce unique products with better taste improved flavor characteristics and/or increased nutritional value
- prevention of oxidation of the product and color stabilization and improvement of the texture of the product such as a higher final bulk volume of the product compared with heated air dehydration or microwave dehydration without osmotic pretreatment (Nijhuis et al., 1996)

Dielectric properties are of primary importance to evaluate the suitability and efficiency of microwave heating of the osmotically pretreated products. Furthermore, dielectric properties give insight in expected heat dissipation, temperature-time profiles, and heating homogeneity. The aim of the present work is therefore:

- study the effect of dielectric properties on heating characteristics
- reveal drying kinetics of the combined osmotic dehydration/microwave drying process

EXPERIMENTAL

Mushrooms (*Agaricus bisporus*), classified as first class, are obtained from local greengrocers. The freshly obtained mushrooms are sorted visually on size and color before processing. The geometry for drying is mushrooms cut in halves. Osmotic treatments are conducted with different processing times, temperatures, and salt concentrations:

- 20°C and 0% NaCl
- 20°C and 15% NaCl
- 45°C and 0% NaCl
- 45°C and 15% NaCl

Processing times are 10, 30, 50, 70, and 110 min. The ratio of material and osmotic solution is 1:5 w/w.

Dry matter content is determined with a dry stove, which is preconditioned at 60°C and the product is predried for 15 h at this temperature. Next the oven temperature is raised to 105°C. After the final 3 h of drying at 105°C, the weighing box is cooled down to room temperature in a dessicator and weighed. Salt content is assessed by grinding the dried samples, dissolving in demineralized water, and after filtration, conductivity of the solution was measured.

Products are microwave dried by putting the product in a 6-kW microwave unit operating at a frequency of 2450 MHz. The waveguide is provided with a power measurement system (HP spinner couplers) that can measure the incident and the reflected power. The product is put on a rotating plate in the vessel, in this way averaging the high- and low-electric field strengths in the cavity, which are resulting from the intrinsic properties of the cavity. Power/load ratio is 4 W/g and the power on time is effectively 20 up to 30% of the total processing time.

Measurement of the dielectric properties is performed with an open-ended coaxial probe connected with a network analyzer. The network analyzer launches a microwave signal down the probe. The signal is reflected from the interface formed by the probe end and the material as a function of the material's dielectric properties. The network analyzer detects the magnitude and phase shift of the reflected signal. The computer calculates the dielectric properties from these data and displays it as a function of frequency.

Temperature profiles are measured during microwave heating with fluor optic thermocouples. All data are recorded automatically by a digital data acquisition system.

RESULTS AND DISCUSSION

DRYING KINETICS OSMOTIC DEHYDRATION

The osmotic dehydration is performed with salt solutions of 10 and 15% with bath temperatures of 20 and 45°C. Figure 19.2 shows that non-osmotic

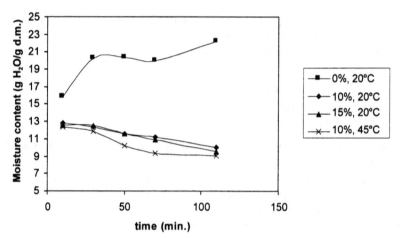

Figure 19.2 Moisture loss during osmotic treatment.

treated mushrooms absorb water as a result of capillary action. The osmotic treatment gives a loss of 30% moisture on initial moisture content. Final moisture content is almost equal for the used osmotic treatments. Higher bath temperatures give a faster decrease in moisture at intermediate processing times. Salt uptake during osmotic treatment is assessed by measuring the conductivity of the solution resulting from grinding and dissolving the osmotic treated mushrooms. Salt gain for the different treatments is shown in Figure 19.3. Higher salt concentrations and temperatures give a salt gain of approximately 0.5 g/g d.m. Salt content of non-osmotic treated mushrooms is very low.

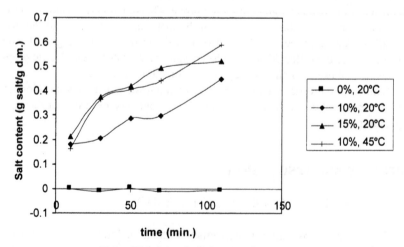

Figure 19.3 Salt gain during osmotic treatment.

DIELECTRIC PROPERTIES OF OSMOTICALLY TREATED PRODUCT

The dielectric properties of the product are described by the dielectric constant ϵ' and the dielectric loss factor ϵ''. These properties determine the suitability of the electromagnetic heating technique. The dielectric loss factor is a measure for the amount of microwave energy that can be dissipated to heat within the product. It also affects the penetration depth of the microwave energy inside the product. A higher loss factor will decrease the penetration depth. Although the absorption of salt by the product results in a higher loss factor, a decrease in moisture content will lower the loss factor. Measurement of dielectric properties for different ratios of salt and moisture within the product show a linear increase in loss factor above a salt content of 0.01 g salt/g available moisture (Figure 19.4). The increase is fivefold from a dielectric loss 12 up to 60. The effect of salt uptake on the increase of the loss factor is much stronger than the decrease by the loss of moisture. The penetration depth of the microwaves at 2450 MHz results in a decrease from 1.1 cm for the raw material to 0.3 cm for the osmotic treated product. A dramatic effect on the heating profiles as a result of the salt uptake can be expected.

The dielectric constant modifies the electric field strength of the microwave field and alters the wavelength inside the product. It describes how much energy is reflected away from the product, how much is transmitted into the product, and is a measure of the concentration of microwaves inside the product. The dielectric constant remains at the same level during osmotic dehydration of approximately 55, so the reflection pattern is basically not changed

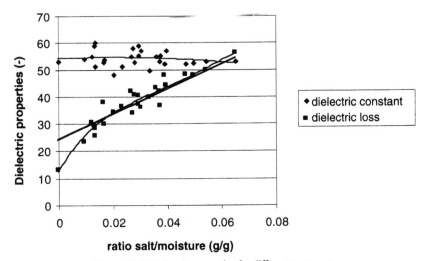

Figure 19.4 Dielectric properties for different treatments.

by the osmotic dehydration process. Generally, from a flat surface the reflection increases with an increasing dielectric constant.

MICROWAVE HEATING CHARACTERISTICS OF OSMOTICALLY TREATED PRODUCT

Heating profiles are measured in the center and surface of the mushroom by insertion of fluoroptic thermocouples. The microwave treatment is performed with a supplied power of 4 W/g.

Measured temperature profiles are shown in Figure 19.5. The center shows a much higher temperature rise for the non-osmotic treated mushroom compared with the surface. The temperature profiles with a higher salt uptake are almost similar for center and surface. The salt uptake results in higher heat dissipation at the surface layers, in this way preventing overheating of the center parts of the concave shaped mushroom. According to the experiments, the shielding effect for overheating of the center parts is effective with ratios salt/moisture in the product above 0.035 g salt/g available moisture (0.3 g salt/g d.m.).

MICROWAVE DRYING KINETICS OF OSMOTICALLY PRE-TREATED PRODUCT

In Figure 19.6 the change in dielectric properties and moisture decrease for the combined process is shown for different power-on times.

With microwave drying, the dielectric properties are rapidly decreasing. The dielectric constant is decreasing as a result of the inclusion of air within the product and the loss factor is decreasing as a result of the loss of moisture.

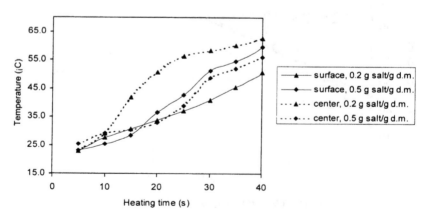

Figure 19.5 Temperature profiles during microwave heating (4 W/g).

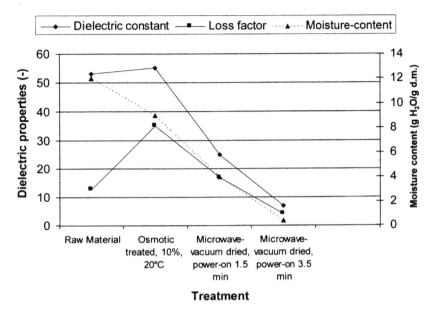

Figure 19.6 Change in product properties during combined treatment.

Moisture content vs. drying time for the different osmotic salt treatments (bath temperature and salt concentration) are shown in Figure 19.7. The drying time is approximately the same for all treatments. It is 1–2 min shorter as a result of the lower moisture content at the beginning of the drying process for osmotically treated mushrooms at higher solute temperatures.

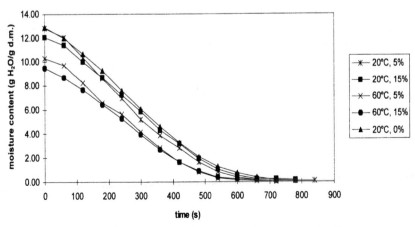

Figure 19.7 Drying curves of mushroom during microwave heating.

CONCLUSIONS

- Osmotic treatment of mushrooms with a salt solution can be dehydrated from 12 -kg moisture until desired salt uptake in 2 h by using salt solutions of 10 up to 15% with temperatures between 20 and 45°C. Moisture content after osmotic dehydration is 9.0 g moisture/g dry matter.
- Dielectric loss increases dramatically with a salt uptake up to a factor 5, so higher heat dissipation within the product can be expected and much lower penetration depth from 1.1 to 0.3 cm is calculated. In this way, surface heating of the product will be intensified. The dielectric constant remains constant, so reflection behavior from the surface will not be modified.
- Microwave heating of the concave shaped mushroom is more homogeneously above salt-moisture ratios of 0.035 g salt/g available moisture.
- Dielectric properties are rapidly decreasing during microwave-vacuum drying
- Drying times for microwave dehydration can be reduced by 10–20% as a result of the lower initial moisture content by osmotic pretreatment

REFERENCES

Lepourhiet, A. and Bories, S. (1980). Analysis of the role of microwave energy contribution in drying porous media. In *Mumjumdar: Drying '80*, 186–194. Hemisphere, New York.

Nijhuis, H. H., Torringa, E., Luyten, H., René, F., Jones, P., Funebo, T., and Ohlsson, T. (1996). Research needs and opportunities in the dry conservation of fruits and vegetables. *Drying Technology*, 14(6):1429–1457.

Nijhuis, H. H., Torringa, H. M., Muresan, S., Yuksel, D., Leguijt, C., and Kloek, W. (1998). Approaches to improving the quality of dried fruit and vegetables. *Trends in Food Science and Technology*, 9:13–20.

Torringa, H. M., Van Dijk, E. J., and Bartels, P. V. (1996). Microwave puffing of vegetables: Modelling and measurements. *Proceedings of the 31st Microwave Power Symposium*, Boston, July 28–31, p. 16–19.

CHAPTER 20

Effect of Vacuum Impregnation on Combined Air-Microwave Drying of Apple

P. FITO
M. E. MARTÍN
N. MARTÍNEZ-NAVARRETE
A. CHIRALT
J. M. CATALÁ
E. DE LOS REYES

INTRODUCTION

MICROWAVE energy is increasingly being used worldwide for the industrial processing of foods. Table 20.1 shows the main applications of microwave energy in the food industry with the major objectives of each operation and some of the food products to be processed. The most successful so far have been tempering, drying, and cooking (George, 1997).

However, in comparison with household appliances, microwave processing in the food industry has not been as successful. The principal reasons for the slow development of microwave energy for industrial processes are related to costs and lack of information about the technology (Schiffmann, 1992; Mudgett, 1986). Also, a new production technique or a new product implies a risk for processors, who are unlikely to displace more established techniques unless huge commercial advantages are anticipated. Use of microwave heating offers many advantages compared with conventional heating methods, in fast operation, energy savings, precise process control, and faster start-up and shutdown times (Decareau, 1985). Because microwaves penetrate within a food product and do not just act at its surface, energy conversion into heat throughout the product is more efficient. This accelerated heating gives a higher quality product in taste, texture, and nutritional content, as well as increased production (Giese, 1992).

The most successful applications of microwave energy have emerged after careful consideration of how and where to apply microwave in the process line (George, 1997). Many successful operations use microwave energy in combination with conventional heating. This can combine the rapid internal

225

TABLE 20.1. Typical Applications for Microwave Energy in Food Processing.

Application	Objectives	Food Products	References
Tempering	Raise temperature below freezing	Meat, fish, butter	—
Vacuum drying		Seed, grain, citrus juices	*
Freeze drying	Reduce moisture content	Meat, vegetables, fruits	**
Drying		Pasta, rice, snack foods	†
Cooking	Modify flavor and texture	Bacon, meat patties, potatoes	—
Blanching	Inactivate spoilage enzymes	Fruit, corn, potatoes	††
Baking	Heating and activating leavening agents	Bread, pastry, doughnuts	—
Roasting	Heating and promoting heat reactions	Nuts, coffee, cocoa	§
Rendering	Melting and separating foods	Bread, dairy products, ready meals	—
Pasteurization	Inactivating vegetative microorganisms	Dairy products, prepared foods	§§
Sterilization	Inactivating microbial spores		—

*Kiranoudis et al., 1997; Pominski et al., 1989.
**Ben Souda et al., 1989; Chen et al., 1993.
†Adu et al., 1996a; Beke et al., 1995; Campaña et al., 1993; Funebo and Ohlsson, 1998; Niewczas et al., 1996; Shivare et al., 1991; Shivare et al., 1994; Walde et al., 1997; Wenzheng et al., 1999.
††Wrolstad, R. E. et al. (1980).
§Fadini, A. L. et al. (1998); Krysiak, W. (1997).
§§Nikdel, S. et al. (1993).
Source: Adapted from George, R. M., 1992.

heating from microwave energy with desirable surface-heating effects from conventional techniques (George, 1997), to produce browning, to facilitate drying, reduce surface bacteria, or preheat the product, depending on the food material (Giese, 1992).

The greatest disadvantage of microwave-heated foods is the non-uniform temperature distributions occurring throughout the bulk of some foods. The control of heating uniformity is of paramount importance to the success of microwave process techniques in the food industry. Care in food engineering can help to minimize these effects (George, 1992). The type of energy distribution system to be employed (multimode cavities or controlled directed heating systems) depends on the application (Ohlsson, 1991).

MICROWAVE BASICS

Microwaves are electromagnetic waves of radiant energy with wavelengths between radio and infrared waves on the electromagnetic spectrum (1 mm–1 m). Microwave frequencies range from 300 MHz to 300 GHz, and microwave heating is defined as the heating of a substance by electromagnetic energy operating in that frequency range (Risman, 1991). To avoid interferences with frequencies employed for communication purposes, only frequencies of 2.45 GHz and 945 MHz are used in food technology applications.

When microwaves interact with materials, their energy content is manifested as heat due to the appearance of several energy conversion mechanisms. The most important when referring to a food material and microwave frequencies are the ionic conduction and the dipolar rotation (Schiffmann, 1995). Both are based on the influence of the electric fields on charge units, which promotes conversion of electric energy into kinetic energy and this into heat.

IONIC CONDUCTION

Ions, as charge units, are made to move in the opposite direction to their own polarity by the electric field, colliding with un-ionized water molecules. Kinetic energy given up to these molecules causes them to accelerate and collide with other water molecules in billiard ball fashion. When the electric field polarity changes, ions accelerate in the opposite way, and because this occurs many millions of times per second, a large number of collisions and energy transfers take place.

DIPOLAR ROTATION

Dipolar molecules, such as water, have an asymmetric charge center, affected by the rapidly changing polarity of the electric field. The dipoles are caused to align continuously with the electric field direction. This occurs many millions of times per second at a frequency, promoting an energy conversion into heat by molecular friction. This type of mechanism is strongly temperature dependent, whereas ionic conduction is not.

It can be noted that the electric component of the electromagnetic waves plays the primary role in heating. Besides, natural biological materials absorb only the electric part of the electromagnetic field, and they do not absorb magnetic field energy (Metaxas and Meredith, 1983; Mudgett, 1986).

DIELECTRIC PROPERTIES

Microwave penetration into food materials depends on the microwave frequency used but is clearly determined by the dielectric properties of the material. Among them, the most significant is the dielectric permittivity, which describes how the food interacts with electromagnetic radiation. This property is a complex quantity, with a real part, the dielectric constant, and an imaginary component, the dielectric loss factor [Equation (1)]. The dielectric constant, ϵ', is a measure of the substance ability to store electrical energy, hence its capacitance; and the dielectric loss tangent, ϵ'', is related to the substance ability to dissipate the energy stored into heat. The ratio of the dielectric loss to dielectric constant is called the loss tangent (tan δ).

$$\epsilon^* = \epsilon' - j\epsilon'' \tag{1}$$

To interpret the meaning of the values of the dielectric properties for foods, a penetration depth can be calculated. This term is defined as the depth into a material where the incident energy of a perpendicularly impinging, forward propagating plane electromagnetic wave on the surface of the material, has decayed to $1/e$ from the surface value ($1/e \approx 37\%$). If ϵ'' is small, the following equation gives a reasonable accuracy for most foods:

$$d_p = \frac{\lambda_o\sqrt{2}}{2\pi} \frac{1}{\sqrt{\epsilon'}\sqrt{1 + (\epsilon''/\epsilon')^2} - 1} \tag{2}$$

where λ_o is the free space wavelength. At a frequency of 2450 MHz, this wavelength is approximately 12.2 cm.

Absorption of microwave energy within a product is determined by an attenuation factor α that is related to the product's dielectric properties ($dp = 1/2\alpha$). The attenuation factor determines the absorption of energy within the dielectric as a function of depth from the surface of the dielectric, as described by Lambert's law [Equation (3)] and is inversely related to the material's penetration depth ($d_p = 1/\alpha$).

$$P_z = P_o e^{-2\alpha z} \tag{3}$$

where P_z is the absorbed energy at z depth from the dielectric surface, P_o is the absorbed energy at the dielectric surface, and α is the attenuation factor.

The dielectric properties are greatly influenced by frequency, temperature, moisture content, salt content, and physical state (Decareau, 1985). Most of the effects of these factors on the dielectric loss may be explained from the lower value of the bound water relaxation frequency. The relaxation time (τ) is the time it takes for an agitated molecule to relax back to 37% of the molecule normal state, and the relaxation frequency (f_r) is related with the relaxation time by $f_r = 2\pi\tau$.

INFLUENCE OF FREQUENCY AND TEMPERATURE ON THE DIELECTRIC PROPERTIES

The temperature dependence of the dielectric constant is quite complex and depends on the material. At submicrowave frequencies in a region of static dielectric behavior, dielectric properties of free water decreases when temperature increases at a fixed frequency. In aqueous ionic solutions, as explained by Decareau (1985), the same effect of temperature on the dielectric loss can be observed when the dipole loss mechanism is the predominant. If the ionic loss effect takes place (at higher temperatures), the dielectric loss increases with rising temperature. For food materials, which are basically complex mixtures of interactive and non-interactive components, the relationship between temperature and dielectric properties at a determined frequency became much more complex. Ohlsson et al. (1974) studied the frequency and temperature dependence of dielectric properties on fat beef, cooked beef, and codfish, raw and cooked potato, mashed potato, thickened gravy, distilled water, and 0.1 M sodium chloride solution. Results revealed that dielectric constant tended to increase and dielectric loss increased significantly with falling frequency, this probably because of the increasing conductivity losses at lower frequencies. A more recent work on fresh fruits and vegetables (Nelson et al, 1994) presents the obtained permittivity values in a frequency range of 0.2–20 GHz. Dielectric constant decreased steadily with increasing frequency, and the loss factor decreased as frequency increased above 200 MHz to a broad minimum, increasing later as the frequency approached 20 GHz. Authors explained this dielectric behavior in predominant ionic conductivity and bound water relaxation at the lower frequencies and predominant free water relaxation at the higher ones.

INFLUENCE OF COMPOSITION ON THE DIELECTRIC PROPERTIES

The influence of water and salt (molecules that are principally affected by dipole and ionic losses) depends on the manner in which they are bound or restricted in their movement by the other food components (Ryynänen, 1995). In a general way, dielectric constant gains with increasing moisture content, and the same path show the dielectric loss, but levels off in the range of 20–30% (Schiffmann, 1995). Water in foods may be in free or bound states depending on the product's water content; the sorption isotherm distinguishes monolayer, multilayer, and capillary condensation regions. Other components, such as salts and sugars, may interact with water molecules at low-moisture contents, leading to tightly bound forms of low dielectric activity.

Mudgett et al. (1980) predicted the dielectric behavior of a semisolid food (potato) at 3 GHz for a range of moisture contents and water activities by an electrophysical model for two-phase heterogeneous mixtures. A critical water content was found for activation of dielectric mechanisms that seemed to be related to the availability of water in mobile forms. Results suggested that bound water relaxation are negligible at microwave frequencies. This appreciation agrees with results reported by Funebo and Ohlsson, (1999), who based the observed peak value of dielectric loss at intermediate moisture contents when measured on several fruits and vegetables at 2.8 GHz on the fact that the reduction of the relaxation frequency as moisture content decreases may lead to an approach to the operating frequency. The dielectric behavior of water, which shows an increase in dielectric loss to a maximum at a critical frequency (relaxation frequency), can be predicted by the Debye equations (Decareau, 1985) for pure solvents as functions of wavelength (or frequency) and temperature.

A low or moderate salt content does not modify the ϵ' values much. Products with a high-salt content (as ham, with 3–4%) present very low penetration depth values. The effect of dissolved salts is to depress the dielectric constant and elevate the dielectric loss with respect to levels observed for water at corresponding temperatures (Hasted et al., 1948). This dielectric behavior has been predicted by the Hasted-Debye models for aqueous ionic solutions (Mudgett et al., 1974a). A recent study on pregelatinized bread with different salt contents (Goedeken, 1997) explained the higher dielectric loss factor values found for samples that contained salt in additional dissipation created by electrical conduction of ionic substances (salt) to dipole-dipole interactions. In this latter work, dielectric constant was not significantly influenced by the presence of salt and both, dielectric constant and dielectric loss factor, increased linearly with respect to mass concentration of water (g_{water}/cm^3_{sample}).

Kent et al. (1987) observed that an increase in sucrose concentration of sugar solutions studied produced a reduction in relaxation frequency. As explained by Decareau (1985), the effect of sugars (as well as other interactive constituents such as alcohols) can be based on hydrogen bonding between hydroxyl groups of these components and water molecules. This effect has been predicted for alcohol-water mixtures by the Maxwell-Debye model for interactive mixtures (Mudgett et al., 1974b). Food products with a high content of dissolved sugars may have a synergistic loss behavior.

The effects of insoluble and immiscible constituents is to depress both the dielectric constant and loss factor in aqueous mixtures as seen in oil-water emulsions (Mudgett et al., 1974b), and this effect is related to lipid, protein, and carbohydrate constituents in "colloidal" suspensions with water (Decareau, 1985). Such behavior has been predicted by Fricke model expressed in terms of complex permittivity (Mudgett et al., 1974b).

INFLUENCE OF BULK DENSITY ON THE DIELECTRIC PROPERTIES

The dielectric properties of foods are determined by their chemical composition, but also, to a much lesser extent, by their physical structure (Ryynänen, 1995). Because the amount of mass interacting with the electromagnetic fields associated with microwaves affect the dielectric behavior, the mass per unit volume (the density), will also have some kind of influence on the dielectric properties. Measurement on particulate materials (pulverized or granular) offers the possibility of analyzing the influence of bulk density on the dielectric properties. As Nelson et al. (1994) remarks in a recent work, the relationship between these variables appeared linear over narrow density ranges of cereal grains and soybeans, but nonlinear behavior has been noted for very wide density ranges. This linearity appears between bulk density and square root and cube root of the dielectric constant of rice (Nelson and Noh, 1992), soybeans (Nelson, 1985a), and hard red winter wheat (Nelson, 1985b). Dielectric mixture equations have been widely used to estimate the dielectric properties of a solid material from the properties of an air-particle mixture made up of air and the pulverized particles of the solid (Nelson, 1992; Nelson et al., 1994; Nelson, 1985a, 1985b). These equations relate the dielectric properties and the bulk density.

In semisolid materials, because dielectric constant of air is equal to 1, a decrease in bulk density may lead to an increase in dielectric properties.

PREDICTIVE EQUATIONS FOR FOOD MATERIALS

All the equations mentioned in previous sections of this work describe the dielectric for model systems. But there is still a lack of equations specific for foods. Those concerning particulate materials are cited before when the influence of bulk density on the dielectric properties is explained. Others are collected in this headline.

Calay et al. (1995) reported several equations to estimate the dielectric properties of a wide range of foods as a function of food composition, moisture content, and temperature. The results obtained may allow to predict the temperature distribution in a food slab subjected to microwave heating.

Another work on vegetables and fruits (potatoes, carrots, apples, peaches, and pears) was conducted by Tran et al. (1984), and a correlation based on Maxwell's mixture theory was found in a frequency range of 0.1–10.0 GHz between dielectric properties and composition.

Mudgett et al. (1974a) confirmed that the nonfat dried milk solutions behave essentially as aqueous ionic solutions; thus, dielectric properties of liquid systems with low colloidal content may be predicted from conductivity measurements with Hasted-Debye models.

RESEARCH ON THE DIELECTRIC PROPERTIES OF APPLE VAR. *GRANNY SMITH*. INFLUENCE OF WATER CONTENT AND VACUUM IMPREGNATION PRETREATMENT

Vacuum impregnation (VI) consists on the partial substitution of internal gas in a porous solid food by an external liquid with a notable reduction of its porosity (Fito, 1994). The hydrodynamic mechanism (HDM) acts as the result of external pressure changes imposed on the liquid-food system (Fito and Pastor, 1994). Because the impregnated food suffers important changes in its physicochemical and structural properties (Chiralt et al., 1999), VI as an operation before drying may result in a useful tool to obtain new products more suitable for microwave drying and so with an improved final quality.

To analyze the influence of VI treatment on the dielectric properties of apple var. *Granny-Smith,* non vacuum impregnated (NVI) and vacuum impregnated (VI) with an isotonic solution (apple juice) samples were prepared as reported in previous work (Martín et al., 1999). VI treatment promoted structural changes in an increase in apparent density (ρ_a) and a decrease of porosity (ϵ) for the VI samples, as a result of the gas release and liquid intake in the food porous structure. The overall concentration of the liquid phase did not change because the impregnation liquid was just an isotonic solution.

Table 20.2 shows the average values obtained in this latter work for the dielectric constant (ϵ'), the dielectric loss (ϵ''), and the loss tangent (tan δ), measured by using an open-ended coaxial probe and a microwave network analyzer at room temperature (20°C) and 2.45 GHz. Values presented are those obtained at high-, intermediate-, and low-moisture contents. Available data on the dielectric properties of some fruits and vegetables for a range of frequencies, measured with the same technique, have been presented in previous articles (Nelson et al., 1994; Tran et al., 1984). As a reference to analyze the influence of water content on the dielectric properties, sucrose-agar gel values are included in the table (Kent and Kress-Rogers, 1987).

Most measured values presented standard deviations below 8%. Only when VI was applied and the samples were highly dehydrated, the standard deviations were higher, probably because of the intensive shrinkage of samples. The measurement technique requires a flat surface of the sample in contact with the probe and a thickness enough to contain the aperture fields within its volume.

Previous work mentioned before represented the complex plane plot (ϵ'' vs. ϵ') of all the dielectric properties values obtained. It could be observed that dielectric loss values of vacuum impregnated samples laid beside those of sucrose-agar gels, evolving with the change of water content in the same way. The values corresponding to non-impregnated samples presented lower val-

TABLE 20.2. Dielectric Properties of Vacuum Impregnated (VI) and Non-Vacuum Impregnated (NVI) Apple and Those of Agar-Sucrose Gels with Different Moisture Content.

Material	X (Wet Basis)	ε'	ε''	Loss Tangent
Dried apple (NVI)	0.12	4.04 ± 0.18	1.0 ± 0.08	0.25
Dried apple (NVI)	0.57	27.8 ± 1.1	10.5 ± 0.3	0.38
Dried apple (NVI)	0.69	38 ± 6	12.0 ± 1.7	0.31
Dried apple (VI)	0.35	1.7 ± 0.7	—	—
Dried apple (VI)	0.43	6 ± 3	4.1 ± 1.2	0.68
Dried apple (VI)	0.62	53.2 ± 1.1	16.3 ± 0.7	0.31
Fresh apple (VI)	0.87	62 ± 6	11.8 ± 1.2	0.19
Fresh apple (NVI)	0.86	56.3 ± 0.9	12.7 ± 0.6	0.23
Water (25°C)*	1.00	76.0	12.0	0.16
Agar-sucrose gel**	0.94	70.7	11.5	0.16
Agar-sucrose gel**	0.90	73.6	11.3	0.15
Agar-sucrose gel**	0.78	70.8	13.4	0.19
Agar-sucrose gel**	0.67	66.8	14.7	0.22
Agar-sucrose gel**	0.45	42.7	15.1	0.35

*Decareau, 1985.
**Kent and Kress-Rogers, 1987.

ues of the loss factor for the same dielectric constant, evolving in a similar manner to those of sucrose-agar gels lying on an ideal parabolic line but with a lower maximum. The corresponding point for pure water also agrees with this behavior. This fact probably reflects the influence of the gas phase in porous fruit (porosity) on their dielectric properties. Because vacuum impregnation implies a sample density increase, due to the controlled substitution of some internal gas by an external liquid, dielectric properties are greatly affected by this treatment. In this kind of samples, as in the sucrose-agar gels, the food liquid phase seems to be the principal responsible for the food-microwaves interaction. Nevertheless, the residual gas present in vacuum impregnated samples (which porosity ranges between 0.05–0.06) will contribute to the differences with the gel sample behavior, which does not contain gas. In the non-impregnated apple samples, the high gas phase volume fraction (0.23) may be responsible for the great differences found in their dielectric properties.

Independent components of this complex property have been analyzed as a function of moisture content in previous work (Martín et al., 2000). It could be observed that VI implied an increase in apple dielectric properties at high water content levels. Below 50% moisture content, an inverse effect was observed. Sucrose agar-gel values overcome those of both VI and NVI apple samples.

DRYING KINETICS

To use microwave energy for industrial drying purposes, it is necessary to understand the heat and mass transfer mechanisms that take place in microwave drying, which must be predictable. Several authors have made an effort to describe microwave drying characteristics, but there is still a lack of basic models for biological materials. As stated in the introduction, microwave heat generation in foods is primarily due to dipolar rotation of free water molecules (Metaxas and Meredith, 1983; Decareau, 1985), and in this sense water binding may play an important role in moisture loss rate along drying process.

Adu and Otten (1996a) investigated the effect of increasing hygroscopicity on the microwave heating of soybeans at 2450 MHz and constant absorbed microwave power input. They observed that the product temperature rise rapidly at the initial stages of the drying process, reaching a maximum, and then decreased gradually with decreasing moisture content during the falling rate period. This behavior revealed that an increase in moisture bond strength promotes an increase of the heat of desorption as the energy needed to free the molecules so they can rotate became higher. Therefore, product hygroscopicity may be considered when modeling microwave drying of foods.

Another work by Adu and Otten (1996b) found that drying characteristics of a thin layer of white beans when heated by microwaves at constant absorbed power were well predicted by the three-term approximation of the theoretical series solution of Fick's diffusion equation for variable diffusion coefficient with time. This seemed to indicate that drying of white beans proceeds mainly in the falling rate period, characteristic of internally controlled diffusion. Shivare et al. (1994) studied the drying kinetics of maize in a microwave environment from an empirical model that substitutes the equilibrium moisture content by the surface moisture value in the analytical solution to Fick's second law of diffusion for a homogeneous and isotropic sphere. This modified equation was also applicable to convective drying of rough rice, microwave drying of wheat, and combined microwave-fluidized bed drying of wheat.

Constant and Moyne (1996) proposed a set of equations to describe the physical mechanisms of heat and mass transfer that occur during drying with internal heat sources, that is, microwaves. They assumed the medium in thermodynamic equilibrium and the total gaseous pressure gradient as an additional driving force (besides the diffusional mechanisms) in both the gas phase and liquid phases when the temperature of the medium exceeds the boiling point imposed by the external pressure.

All the mentioned works assumed that microwave power absorption was uniform throughout the food material at any given instant of drying time, so mass transfer could be described more easily. Nevertheless, temperature profiles may be also studied when developing models that try to reproduce the microwave drying kinetics. The finite element method (FEM) is a relative new numerical method employed in food engineering, but its flexibility in handling irregular geometrical configurations and material properties and its accuracy in analyzing non-homogeneous and non-isotropic food products make it the preferred numerical method to solve the microwave heating process in foods (Lin et al., 1995). Moreover, temperature and moisture distribution in food materials can be predicted from this method in most kind of foods during microwave heating. Zhou et al. (1994a) investigated the effect of temperature gradient on moisture gradient during microwave heating of potato and bread by using the finite element method, and they found that for high density food (potato), this effect was negligible, whereas for low density and porous food material (bread), most moisture movement was due to temperature gradient. A three-dimensional finite element model was developed and validated by Zhou et al. (1994b) with analytical solutions, commercial software (Twodepep), and experimental data of cylinder and slab-shaped potato. The FEM-predicted temperature agreed with measured results. Absorbed microwave density at any location in sodium alginate gel was derived as a function of dielectric properties and geometry (rectangular and cylindrical) of the material by Lin et al. (1995) by using a two-dimensional commercial finite element software (Twodepep). They found that temperature predictions in slab-shaped samples agreed with experimental measurements, but for cylindrical samples, they deviate in the central region. Moreover, this work explains how a variation in thermal diffusivity, dielectric properties, or incident microwave power resulted in significant variation in the predicted temperatures. More recent works by Vilayannur et al. (1998a, 1998b) analyze the influence of the size and shape effect on non-uniformity of temperature and moisture distributions in microwave-heated potato samples by using the element finite method.

Other authors have made an effort to predict temperature distribution throughout the food material from Maxwell's heat equations for microwave energy distribution. Fleischman et al. (1999) used this integral transform technique over raw beef, and results revealed a significant temperature range sensitivity to slab width, which may have clear implications for food service operations where pans of food are heated by microwaves.

Combined air-microwave drying has been studied to a much lesser extent. As Funebo and Ohlsson (1999) explains, a combination of hot air and microwave energy improves the heat transfer compared with hot air alone, leading to better aroma, better and faster rehydration, and much shorter drying times. Shivare et al. (1991) dried seed grade corn in a modified microwave oven by using an incident microwave power range from 60 to 600 W and a cross flow of hot air (30, 50, and 80°C). The total drying time for corn could be reduced considerably by increasing the microwave power input and the inlet air temperature. Results also revealed that the thin layer concept can only be applied for the beds less than 0.05 m in high as a temperature gradient across the grain bed was found while combined drying. Prabhanjan et al. (1995) studied the microwave-assisted air drying of thin layer carrots in a domestic microwave oven at two incident microwave power levels and two air temperatures (45 and 60°C). The drying time was reduced 25–90%, and the air temperature influence on drying rates seemed to be clearly diminished at the higher power levels due to microwave-facilitated outward flux of water vapor. They also found that diffusion models did not fit well the experimental results, and they proposed the empirical Page's model. Microwave-assisted air dehydration of apple and mushroom was investigated by Funebo and Ohlsson (1999) at low power microwave energy (0.6–1 W/g food), three air rates (0.5, 1, and 1.5 m/s) and 40, 60, and 80°C air temperature. They used the internal food temperature as an indicator of the absorbed power, which remained constant along the drying process. As expected, air velocity, absorbed microwave power, and air temperature were inversely proportional to the drying time, being the center temperature of the food more important than the air temperature. In this latter work, rehydration capacity, bulk density, and color of dried products were tested.

A PRELIMINARY STUDY ON THE COMBINED MICROWAVE-AIR DRYING (MW-AD) OF APPLE VAR. *GRANNY SMITH*: INFLUENCE OF VACUUM IMPREGNATION (VI) PRETREATMENT

To analyze the effect of microwave energy application on the air-drying kinetics of apple, different experimental conditions were conducted in a laboratory microwave-convective dryer. The effect of VI with an isotonic solution as a treatment before drying was also studied. Experimental conditions were 40 and 50°C for air temperature, 2 m/s for air rate, and incident microwave power ranging between 0 and 4 W/g sample.

Laboratory equipment was designed and constructed as described in previous work (Martín et al., 1999). It allows to apply hot air and/or microwave energy into a multimode cavity, where the sample is suspended from a top mounted balance. Air temperature, room temperature, humidity, and velocity, as well as the incident and reflected microwave power, were controlled and recorded with a computer. Changes in sample weight throughout the drying process were also monitorized.

As expected, the application of MW implies a dramatic decrease in the drying time. It may also be observed, as a first step in the kinetics study, that VI pretreatment promote drying rate only when MW are applied.

To establish mathematical equations that allow to predict apple drying behavior in the studied range, drying curves were plotted as logarithm of reduced water content (dry basis) as a function of time for all the treatments. Over moisture content levels of 2 g/g dry solids, linear relationships between plotted variables could be found. As deduced from this linear behavior, Equation (4) could be accepted to reproduce obtained results. This suggests that a first-order kinetics equation for water loss could be assumed [Equation (5)] to model drying, empirically. The only parameter (kinetic constant: k) that appears in Equaton (5) can be obtained from fitting the corresponding linear equations [Equation (6)] to experimental points (Martín et al., 2000). Table 20.3 shows the obtained values of k for each drying condition. Kinetic constant is related to water flux according to Equation (7).

$$X_w = X_{w_0} e^{-k \cdot t} \tag{4}$$

$$-\frac{dX_w}{dt} = kX_w \tag{5}$$

$$\ln X_w = \ln X_{w_0} - kt \tag{6}$$

$$J_w = -\frac{1}{A}\frac{dX_w}{dt} = k\frac{X_w}{A} \tag{7}$$

where:

J_w: water flux (g/m^2 s)
A: evaporation surface (m^2)
X_w: moisture content (d.b.)
k: drying constant rate (s^{-1})

Results concerning to MW-AD drying kinetics of VI and NVI apple, var. *Granny Smith*, in the range of treatment conditions presented in this work, may define the incident microwave power as the main factor responsible for

TABLE 20.3. Model Parameter Obtained for the Different Conditions.

T (°C)	VI/NVI	P_i (W/g sample)	$k \cdot 10^4$ (s^{-1})
50	VI	0.0	1.1
		1.6	5.6
		3.7	15.7
	NVI	0.0	1.6
		1.6	5.4
		2.8	14.2
40	VI	0.0	0.6
		1.6	5.2
		3.0	11.4
	NVI	0.0	1.2
		1.8	6.3
		3.2	12.2

the drying rate. VI pretreatment and air temperature did not seem to have any relevant influence in the process when microwave energy is applied. However, a deeper study in this field is necessary to reach clear conclusions.

REFERENCES

Adu, B. and Otten, L. 1996a. Effect of increasing hygroscopicity on the microwave heating of solid foods. *Journal of Food Engineering,* 27:35–44.

Adu, B. and Otten, L. 1996b. Diffusion characteristics of white beans during microwave drying. *Journal of Agricultural Engineering Research,* 64:61–70.

Beke, J., Mujumdar, A. S., and Bosisio, R. G. 1995. Drying of fresh and rewetted shelled corn in microwave fields. *Drying Technology,* 13(1, 2):463–475.

Ben Souda, K., Akyel, C., and Bilgen, E. 1989. Freeze dehydration of milk using microwave energy. *International Microwave Power Institute,* 24(4):195–202.

Calay, R. K., Newborough, M., Probert, D., and Calay, P. S. 1995. Predictive equations for the dielectric properties of foods. *International Journal of Food Science and Technology,* 29:699–713.

Campaña, L. E., Sempé, M. E., and Filgueira, R. R. 1993. Physical, chemical, and baking properties of wheat dried with microwave energy. *Cereal Chemistry,* 70(6):760–762.

Chen, S. D., Ofoli, R. Y., Scott, E. P., and Asmussen, J. 1993. Volatile retention in microwave freeze-dried model foods. *Journal of Food Science,* 58(5):1157–1161.

Chiralt, A., Fito, P., Andrés, A., Barat, J. M., Martínez-Monzó, J., and Martínez-Navarrete, N. 1999. Vacuum impregnation: A tool in minimally processing of foods. In: Oliveira, F. A. R., Oliveira, J. C., eds. *Processing of Foods: Quality Optimization and Process Assesment.* Boca Ratón, CRC Press; pp. 341–356.

Constant, T. and Moyne, C. 1996. Drying with internal heat generation: theoretical aspects and application to microwave heating. *AIChE Journal,* 42(2):359–368.

Decareau, R. V. 1985. *Microwaves in the Food Processing Industry.* Academic Press, New York.

Fadini, A. L., Gilabert, M. V., Pezoa, N. H., and Marsaioli, Jr. 1998. A study of a continuous roasting process for cocoa using microwaves. *Proceedings of the 11th International Drying Symposium.*

Fito, P. 1994. Modelling of vacuum osmotic dehydration of food. *Journal of Food Engineering,* 22:313–328.

Fito, P. and Pastor, R. 1994. On some non-diffusional mechanism occurring during vacuum osmotic dehydration. *Journal of Food Engineering.,* 21:513–519.

Fleischman, G. J. 1999. Predicting temperature range in food slabs undergoing short-term/high-power microwave heating. *Journal of Food Engineering,* 40:81–88.

Funebo, T. and Ohlsson, T. 1998. Microwave-assisted air dehydration of apple and mushroom. *Journal of Food Engineering,* 38:353–367.

Funebo, T. and Ohlsson, T. 1999. Dielectric properties of fruits and vegetables as a function of temperature and moisture content. *Journal of Microwave Power and Electromagnetic Energy,* 34(1):42–54.

George, R. M. August 1992. A review of the applications of microwave energy in food processing. Technical Bulletin No. 89. CAMPDEN Food & Drink Research Association.

George, M. July 1997. Industrial microwave food processing. *International Review— Food Review*, 11–13.

Giese, J. 1992. Advances in microwave food processing. *Food Technology*, September: 118–123.

Goedeken, D. L., Tong, C. H., and Virtanen, A. J. 1997. Dielectric properties of a prege-latinized bread system at 2450 Mhz as a function of temperature, moisture, salt and specific volume. *Journal of Food Science*, 62(1):145–149.

Hasted, J. T., Ritson, D. M., and Collie, C. H. 1948. Dielectric properties of aqueous ionic solutions. Parts 1 and 2. *J. Chem. Phys*, 16:1–21.

Kent, M., Kress-Rogers, E. 1987. The COST90bis collaborative work on the di-electric properties of foods. In: Jowitt, R., Escher, F., Kent, M., Mckenna, B., Roques, M. *Physical Porperties of Foods—2*. London, Elsevier Applied Science pp. 171–197.

Kiranoudis, T., Tsami, E., and Maroulis, Z. B. 1997. Microwave vacuum drying ki-netics of some fruits. *Drying Technology*, 15(10):2421–2440.

Krysiak, W. 1997. Applications of high frequency waves in roasting cocoa bean. In: *Properties of Water in Foods*, pp. 72–88.

Lin, Y. E., Anantheswaran, R. C., and Puri, V. M. 1995. Finite element analysis of mi-crowave heating of solid foods. *Journal of Food Engineering*, 25:85–112.

Martín, M. E., Fito, P., Martínez-Navarrete, N., and Chiralt, A. 1999. Combined air-microwave drying of fruit as affected by vacuum impregnation treatments. *Pro-ceedings of the 6th Conference of Food Engineering*, pp. 465–470.

Martín, M. E., Andrés, A., Martínez-Navarrete, N., Chiralt, A., and Fito, P. 2000. Di-electric properties of apple Granny-Smith as affected by vacuum impregnation treat-ment and water content. *International Congress of Engineering of Foods* (Icef8). CD-rom support.

Metaxas, A. C. and Meredith, R. J. 1983. *Industrial Microwave Heating*. Peter Pere-grinus, Ltd.

Mudgett, R. E. 1986. Microwave properties and heating characteristics of foods. *Food Technology*, 40(6):84–93, 98.

Mudgett, R. E., Wang, D. I. C., and Goldblith, S. A. 1974b. Prediction of dielectric properties in oil-water and alcohol water mixtures at 3000 MHz, 25°C based on pure component properties. *Journal of Food Science*, 39(3):632–635.

Mudgett, R. E., Goldblith, S. A., Wang, D. I. C., and Wetphal, W. B. 1980. Dielectric behavior of a semi-solid food at low, intermediate and high moisture contents. *Jour-nal of Microwave Power*, 15(1):27–36.

Mudgett, R. E., Smith, A. C., Wang, D. I. C., and Goldblith, S. A. 1974a. Prediction of dielectric properties in nonfat milk at frequencies and temperatures of interest in microwave processing. *Journal of Food Science*, 39(1):52–54.

Nelson, S., Forbus, W. Jr., and Lawrence, K. 1994. Permittivities of fresh fruits and vegetables at 0.2 to 20 GHz. *Journal of Microwave Power and Electromagnetic En-ergy*, 29(2):81–93.

Nelson, S. O. 1985a. A model for estimating the dielectric constant of soybeans. *Trans-actions of the ASAE*, 28(6):2047–2050.

Nelson, S. O. 1985b. A mathematical model for estimating the dielectric constant of hard red winter wheat. *Transactions of the ASAE*, pp. 234–238.

Nelson, S. O. 1992. Correlating dielectric properties of solids and particulate samples through mixture relationships. *Transactions of the ASAE*, 35(2):625–629.

Nelson, S. O. and Noh, S. H. 1992. Mathematical models for the dielectric constants of rice. *Transactions of the ASAE,* 35(5):1533–1536.

Niewczas, J., Wozniak, W., and Kudra, T. 1996. Effect of some microwave drying parameters on the conditions of wheat grain endosperm. In: *Properties of Water in Foods,* pp. 77–85.

Nikdel, S., Chen, C. S., Parish, M. E., MacKellar, D. G., and Friedrich, L. M. 1993. Pasteurization of citrus juice with microwave energy in a continuous-flow unit. *Journal of Agricultural Food Chemistry,* 41:2116–2119.

Ohlsson, T., Bengtsson, E., and Risman, P. O. 1974. The frequency and temperature dependence of dielectric food data as determined by a cavity perturbation technique. *Journal of Microwave Power,* 9(2).

Ohlsson, T. 1991. Microwave processing in the food industry. *E.F. + D.R. European Food and Drink Review.* Spring: 7.

Pominski, J. and Vinnett, C. H. 1989. Production of peanut flour from microwave vacuum-dried peanuts. *Journal of Food Science,* 54(1):187–189.

Prabhanjan, D. G., Ramaswamy, H. S., and Raghavan, G. S. V. 1995. Microwave-assisted convective air drying of thin layer carrots. *Journal of Food Engineering,* 25:283–293.

Risman, P. O. 1991. Terminology and notation of microwave power and electromagnetic energy. *Journal of Microwave Power and Electromagnetic Energy,* 26:409–429.

Ryynänen, S. 1995. The electromagnetic properties of food materials: a review of the basic principles. *Journal of Food Engineering,* 26:409–429.

Schiffmann, R. F. 1992. Microwave food processing: Past, present, and future. Paper 148, presented at *52nd Annual Meeting of Inst. of Food Technologists,* New Orleans, LA, June 21–24.

Schiffmann, R. F. 1995. Microwave and dielectric drying. In: Mujumdar, A. S. *Handbook of Industrial Drying,* Vol. 1. New York: Marcel Dekker, Inc.: pp. 345–372.

Shivare, U. S., Raghavan, G. S. V., and Bosisio, R. G. 1994. Modelling the drying kinetics of maize in a microwave environment. *Journal of Agricultural Engineering Research,* 57:199–205.

Shivhare, U. S., Raghavan, G. S. V., Kudra, T., Mujumdar, A. S., and van de Voort, F. R. 1991. Through-circulation microwave drying of corn. In: *Proceedings of the International Drying Symposium 1991,* pp. 414–421, edited by Elsevier Science Publishers, Amsterdam.

Tran, V. N., Stuchly, S. S., and Kraszewski, A. 1984. Dielectric properties of selected vegetables and fruits at 0.1–10.0 GHz. *Journal of Microwave Power,* 19(4):251–258.

Vilayannur, R. S., Puri, V. M., and Anantheswaran, R. C. 1998a. Temperature and moisture distributions in microwave heated food materials: Part I Simulation. *Journal of Food Process Engineering,* 21:209–233.

Vilayannur, R. S., Puri, V. M., and Anantheswaran, R. C. 1998b. Size and shape effect on nonuniformity of temperature and moisture distributions in microwave heated food materials: Part II. Experimental validation. *Journal of Food Process Engineering,* 21:235–248.

Walde, S. G., Balaswamy, R. S., Chakkaravarthi, A., and Rao, D. G. 1997. Microwave drying and grinding characteristics of Gum Karaya (*Sterculia urens*). *Journal of Food Engineering,* 31:305–313.

Wenzheng, C., Yuhvang, Y., and Zhizhang, C. 1999. Microwave drying of foods with high humidity. *Microwave and Optical Technical Letters,* 22(3):205–207.

Wrolstad, R. E., Lee, D. D., and Poei, M. S. 1980. Effect of microwave blanching on the color and composition of strawberry concentrate. *Journal of Food Science,* 45:1573–1577.

Zhou, L., Puri, V. M., and Anantheswaran, R. C. 1994a. Effect of temperature gradient on moisture migration during microwave heating. *Drying Technology,* 12(4):777–798.

Zhou, L., Puri, V. M., Anantheswaran, R. C., and Yeh, G. 1994b. Finite element modeling of heat and mass transfer in food materials during microwave heating—Model development and validation. *Journal of Food Engineering,* 25(4):509–529.

Index

Milton Keynes UK
Ingram Content Group UK Ltd.
UKHW020025071024
449327UK00032B/2925

9 780367 455248